THE FLOWERING OF GONDWANA

THE FLOWERING

PRINCETON UNIVERSITY PRESS
PRINCETON, NEW JERSEY

MARY E. WHITE
PHOTOGRAPHY OF FOSSILS BY JIM FRAZIER

OF GONDWANA

ACKNOWLEDGEMENTS

The author acknowledges with gratitude her indebtedness to the following:

To Jim Frazier, for his skill and artistry in photographing the fossils (no one could have done it better) and for his help and encouragement with the whole project.

For loan of fossil specimens photographed to illustrate *The Flowering of Gondwana*:
The Australian Museum, Sydney, with kind permission of the Trustees
The Bureau of Mineral Resources, Canberra
The Mining Museum, Sydney
The Museum of Victoria, Melbourne
Dr Jack Douglas, Mines Department, Melbourne
Mr Ron Smythe, private collector, Hobart

The following generously supplied photographs:
Mantis Wildlife Films Pty Ltd whose library of photographs was made available to supply the contemporary component of the illustrations. The generosity of Densey Clyne, allowing the use of her beautiful plant and landscape photographs, and those taken for Mantis by Glen Carruthers and Jim Frazier, has helped to make the book beautiful, and I am deeply grateful.
Dr M. R. Walter of the Baasbecking Institute, BMR, Canberra for the photographs of Stromatolites (including underwater landscape taken for him by Steve Parker)
Dr P. R. Evans, University of New South Wales, for spore photographs
Dr Peter Valder, University of Sydney, for photographs of some rare plants
Dr Alex Ritchie, The Australian Museum, for Antarctic landscapes
Dr John Webb, Melbourne University, for fossil Eucalypt photographs
Dr Peter Jell, Museum of Victoria, for Insect photograph
Mr David Barnes, Photographer, Dept. of Mines, Sydney, for his photograph of Sporangia

The contribution made by Dr Chris Scotese of Texas University in giving me permission to use his continental reassemblies and Gondwanan Break-up sequence is gratefully acknowledged.

My special thanks go to H. F. Doutch, Dr Elizabeth Truswell, G. Wilford, J. Totterdell and others at the BMR, Canberra, who have helped and advised me; to Dr Peter Valder of Sydney University and Dr P. R. Evans of the University of New South Wales; to Dr Alex Ritchie of the Australian Museum and Dr John Pickett of the Mining Museum, Sydney; and to friends and colleagues too numerous to mention individually who have answered questions, corrected my mistakes and given me encouragement and help.

To the team who have turned the manuscript into a book, I extend my special thanks:
For making the maps, Heike Apps in Canberra; for assistance in preparing a simplified geological map, Erwin Feeken; for the illustrations, Dorothy Dunphy; for editing, Helen Grasswill; for the design of the book and making the charts and diagrams, Bruno Grasswill, whose contribution to the final product is beyond calculation. Each has contributed unique skills, and their dedication and enthusiasm was an inspiration.

To Bill Templeman, Publishing Director of Reed Books, Sydney, for bringing *The Flowering of Gondwana* to fruition, for his interest and assistance, my sincere appreciation and gratitude.

ENDPAPERS
Cladophlebis australis, a fossil fern, from Triassic at Kingaroy, Qld., Age about 225 million years.

PREVIOUS PAGE
1 A Honeyeater feeding on *Eucalyptus* blossoms. (Jim Frazier)

2 *Stylidium graminifolium*, the Trigger Plant, an Australian endemic species of a genus which has a Gondwanan distribution pattern. (Densey Clyne)

Published by Princeton University Press,
41 William Street, Princeton, New Jersey 08540
By arrangement with Reed Books Pty. Ltd., Australia

Library of Congress Cataloging-in Publication Data

White, M. E. (Mary E.)
[Greening of Gondwana]
The flowering of Gondwana / Mary E. White.
p. cm.
Reprint. Originally published: The greening of Gondwana. Frenchs Forest, NSW: Reed, 1986.
Includes bibliographical references.
ISBN 0-691-08592-7 : $49.50
1. Paleobotany – Australia. 2. Botany – Australia. 3. Gondwana (Geology) I. Title.
QE948.A1W47 1990 90-31732
581.994–dc20 CIP

Produced in Australia by the Publisher
Edited by Helen Grasswill
Designed by Bruno Grasswill
Fossil photography by Jim Frazier. Most photography of living plants and landscapes by Densey Clyne, Mantis Wildlife Films Pty Ltd.
Cartography by Heike Apps
Illustrations by Dorothy Dunphy
Proofreading by Rosemary Frazer

Typeset in Australia by Deblaere Typesetting Pty Ltd
Printed in Singapore

This book is dedicated to the memory of Bill White, Geologist (1923-1981).

PREFACE

Modern Man is obsessed with time. Watches, clocks and calendars organise our lives. We think in lifetimes. Antiquity to us may be our grandparents' generation, or the birth of Christ, or civilisations BC. We prepare to celebrate two hundred years of the life of our nation with fanfares and festivals, and derive some satisfaction in having so much history. Yet all the days of Man on this Earth of ours, let alone of white man in Australia, pass in the wink of an eye when measured against the Clock of Life.

If we put the present into perspective and think in geological time, in eons instead of generations, we realise we have a heritage so ancient that it disappears into the very mists of creation. Here in our ancient land we have evidence of the oldest rocks formed in the first cooling of the Earth's crust. We have the oldest fossil evidence of life yet found. And some of the Australian landforms are so little changed that when we photograph them now we are recording them much as they were 200 million years ago. We can stand in awed silence in Rainforests whose ancestors were here 60 million years ago, and feel their mystical antiquity, and be humbled in the knowledge that their green, lovely world predates the arrival of humans on this planet by about 58 million years.

Yet even Rainforest is a newcomer in the annals of this country's flora. Fossils record 400 million years of evolution of land-plants in the rocks which are part of the land surface that is now Australia.

The presentation of this account of the vegetation of Australia is intended to fulfil a number of objectives. Firstly, it illustrates a representative selection of Australian fossils and traces the evolution of land-plants from their first appearance, through the geological ages, to the present day. Secondly, it shows the vegetation of each age in the context of the evolving world in which it lived — in an environmental and a palaeogeographical perspective. Thirdly, it shows that Australia as we know it, the island-continent, only became a separate entity in the recent geological past. And finally, it analyses the modern Australian flora, showing how it

3

4

PREVIOUS PAGE

3 Rainforest, with Tree-ferns. Relict Rainforests are a little changed part of Australia's Gondwanan inheritance. Millaa Millaa Falls, North Queensland. (Densey Clyne)

4 Saltbush scrub, a vegetation type often growing on the margins of deserts. It provides surprisingly good grazing for hordes of sheep, which have largely ousted the kangaroos whose territory it was before the sheep were introduced. (Jim Frazier)

evolved out of the ancestral floras which we know from the Fossil Record and thus accounting for the similarities to and differences from other world floras of today.

Above all, this book is intended as a celebration of the 400 million years of the Flora of Australia.

It is necessary to see this 400 million years in context. Although it seems impossible to visualise so much time, we are dealing with only the top ninth of a Time Column which started 3500 million years ago, with the first fossil evidence of life. That this oldest evidence has been found in Australia is further reason for our interest.

To understand this long history of our flora, and our world, we must learn to think in millions of years, not in decades and centuries. Like an astronaut who thinks in millions of kilometres as he travels to the Moon, and as he sees our Earth from outer space, in the context of the Universe, we must try to appreciate the magnitude of time before the present, and see our brief lifetimes against a background of past ages.

To try to tell the story of 400 million years of evolution of the Australian flora is a daunting task. So much to tell, so much which needs to be explained — for the plants were only part of the changing and evolving world.

Where to begin?

Remembering the good advice given to Alice-in-Wonderland by the Red Queen — "start at the beginning and go on until you come to the end" — *The Greening Of Gondwana* starts at the *very* beginning.

Mary E. White
Sydney, Australia
April 1986

CLOCK OF LIFE

To give some perspective of the great age of our Earth, geological time is often compared with a calendar of the year, or with a clock cycle. In relative terms, life on the land would have started in late November on our calendar and it would have been here for only an hour and a quarter if measuring it on our clock.

In terms of a full year, Modern Man appeared on the scene at a minute to midnight on the 31st of December.

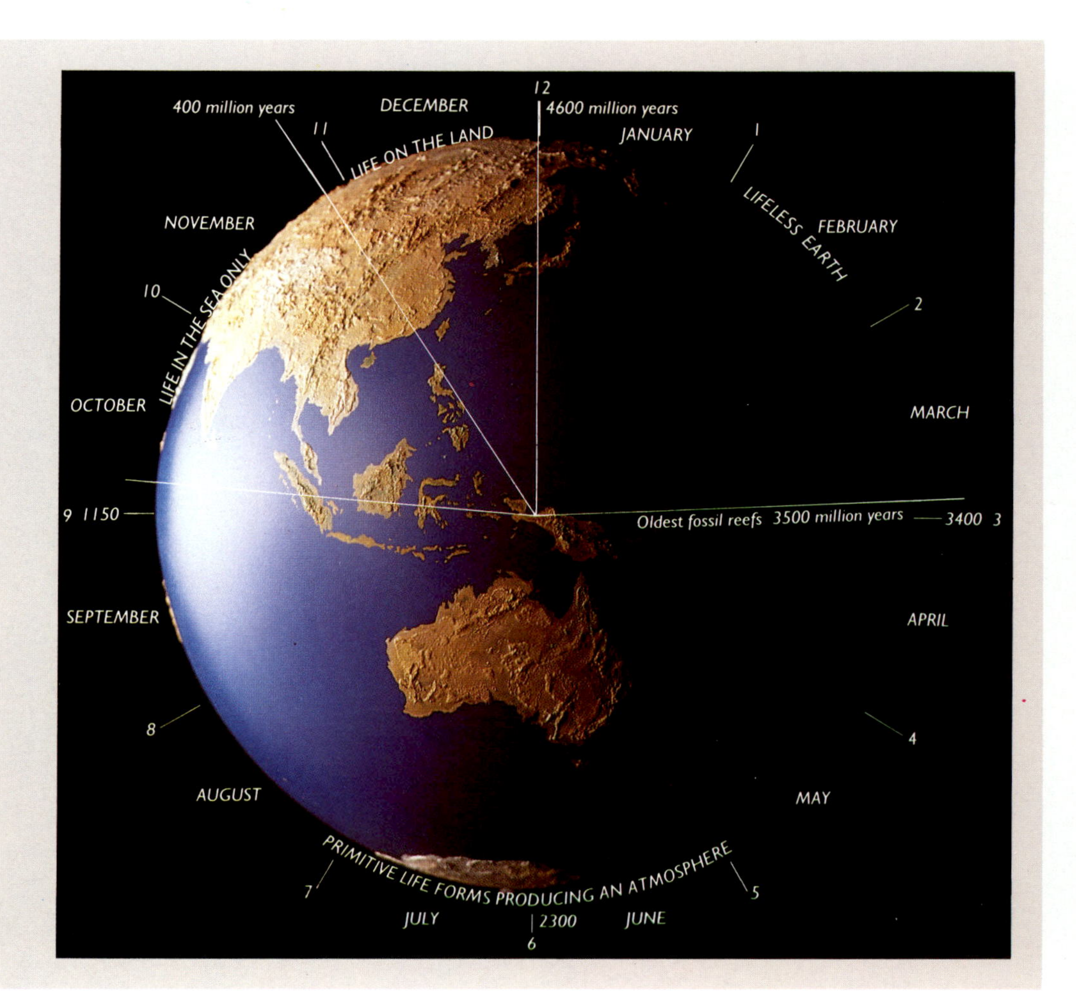

GENERAL INFORMATION

The **time scale** used in this book follows the 1983 Geologic Time Scale in the *Decade of North American Geology* produced by the Geological Society of America and published in *Geology*, September 1983, page 504. It was chosen because it is the time scale used in the continental reassemblies by Dr C. Scotese, which are used in this book with his kind permission. Van den Grinten projections are used.

The series of **palaeogeographic maps** and descriptions of geology of each Period which show the evolution of the Australian continent are based on the "Earth Science Series" of the Bureau of Mineral Resources, Canberra, compiled by G. Wilford, with permission. The north-south orientation of the continent shown in this book follows the palaeomagnetic information for Australia used in compiling the earth science maps. There is some discrepancy between these orientations and those shown in the Scotese reassemblies, due to the bias towards reliable information for the Northern Hemisphere and different information from Australia in the latter. Reassemblies are for an Earth of the same size as now. Dr Scotese does not subscribe to an expanding Earth theory.

Palaeogeographic maps are used as **locality maps for plant fossils** described and illustrated. In descriptions of the palaeogeographic maps, the points of the compass are used as they refer to the modern continent, i.e. if the seas are said to "retreat to the east" this is the east as it is now.

KEY TO PALAEOGEOGRAPHIC MAPS

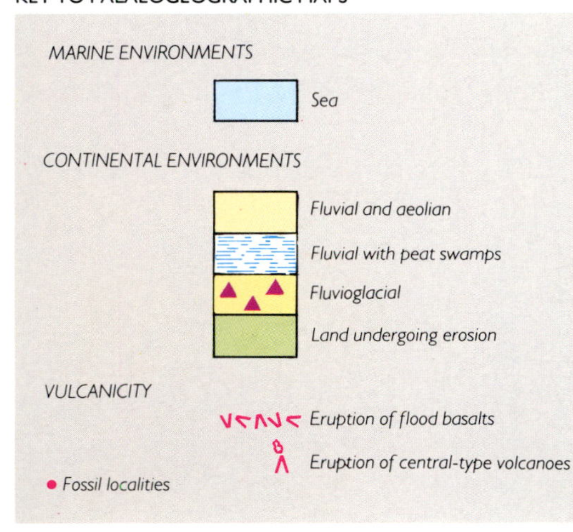

References to literature are indicated (1-123) in margins adjacent to the information to which they apply. A numbered scientific Bibliography is on pages 244-246.

A **Glossary** on pages 241-243 explains terms and words which may not be familiar to all readers.

The **Fossil Specimens** illustrated are listed with information on the collection they are from, locality, age, and number of illustration(s) in an appendix on pages 247-252. Wherever possible, the magnification has been included (between brackets) at the end of captions.

CONTENTS

5 *Triassic plants from Benolong, New South Wales: Dicroidium odontopteroides* (top) and *Dicroidium elongatum* (below). Age about 225 million years.

INTRODUCTION

This book is a window through which we can look down the corridors of time to distant ages and even to the very beginnings when our old Earth was new and first life was born. We can focus on the time when microscopic organisms, in their multitudinous millions in the primaeval oceans, produced an atmosphere suitable for life as we know it to breathe, and started the processes of evolution of plants and animals. It comes as a surprise to realise that Cyanobacteria predestined the course which all subsequent life was to take on our planet.

From such humble beginnings we can trace the evolution of plants from the times when all life was confined to the waters and the land was barren, to the emergence of life on to the land about 400 million years ago, and all through the ages and stages right up to the present. We will see the transformation of a drab and colourless world into a green and lovely place finally crowned with flowers, and understand a little of the magic and the mystery of creation.

The Fossil Record, which when interpreted provides the window into the past, is not just a collection of dusty rock specimens with remains of long-dead plants in their substance. It is a tantalising key to visualising the world as it was long, long ago. The plants tell us of past climates, they conjure up a picture of ancient landscapes, they explain configurations of land and sea in the ever-changing world of the past. When added to our knowledge of geology, of ancient geography and climates, and information about fossil fauna, we can recreate the environments of the land which was to be Australia at different stages of its history.

Through our window into the past we can see the changing face of the globe as parts of the Earth's crust move, and land and sea are rearranged. The Australian land surface was part of larger landmasses for most of its history — first of Pangaea, when all the continents were aggregated into one supercontinent, and later of the Great Southern Continent, Gondwana, when Pangaea had separated into northern and southern regions. It became the island-continent only when it severed its final links with other southern lands about 45 million years ago. Its movements on the face of the Earth while part of other landmasses and later while a separate entity make a fascinating story: from north of the Equator to the South Pole and everywhere between, rotating; no "*terra firma*", this *Terra Australis* — and carrying in it an ever-changing population of plants and animals.

Thus our fossil flora is in large part the flora of "Australia-in-Gondwana", and only in the relatively recent geological past is it the fossil flora of the island-continent. Having fossil floras which form a complete sequence through the geological periods is our great good fortune. It is a reflection of the very old and stable land surface which has become Australia.

The modern flora, the subject of so many books of beautiful illustrations, is "uniquely Australian". When analysed it shows its Gondwanan links, for its ancestors were here while Australia was part of the Great Southern Continent. Our Rainforests are "museums" of relict Gondwanan plants. The "uniquely Australian" element has evolved here in isolation from ancestral Gondwanan stock.

6 Petrophile linearis, Family PROTEACEAE. *The genus Petrophile is exclusively Australian; the family to which it belongs is an ancient Gondwanan one. (Densey Clyne)*

PART ONE

AUSTRALIA'S GEOLOGICAL BACKGROUND AND GONDWANAN INHERITANCE

TIME COLUMN

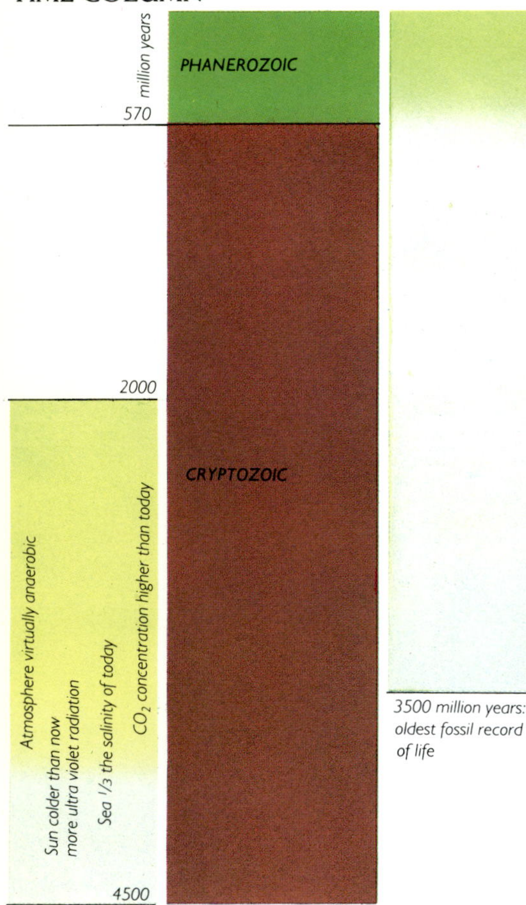

FROM LIFELESS EARTH TO THE AGE OF HIDDEN LIFE

Planet Earth was probably formed about 4600 million years ago. The enormous time span taken to transform the dead planet into the living Earth of today is called "geological time". It is divided primarily into two parts, the Cryptozoic Eon and the Phanerozoic Eon. The names are derived from Greek words, Kryptos *(meaning "hidden"),* Phaneros *(meaning "evident") and* Zoe *(meaning "life"). So, the Cryptozoic Eon is not characterised by living things, but the Phanerozoic Eon is. . . .*

As the Earth cooled from its fiery, molten state at the beginning of the Cryptozoic Eon, a crust formed. Water vapour condensed from the swirling clouds of gases around it, accumulating in the hollows of the surface and forming the first oceans.

No one can hope to know the configurations of the first land and seas. Violent instability would have been a characteristic of the young planet's land surfaces.

The Cryptozoic Eon embraces more than 80 per cent of geological time. Its rocks are known as Precambrian rocks, and they were formed before life, as we know it, was apparent on Earth. This major segment of the history of the Earth lasted for 4000 million years or more, and we need to know a little about it in order to gain a perspective of the background to our fossil flora story.

THE ROCKS OF THE EARTH'S CRUST

PREVIOUS PAGE
7 Ephemerals blooming in the desert among the "spinifex". The Olgas, Central Australia. *(Densey Clyne)*

The rocks of the original crust and subsequent rocks produced by volcanic eruptions from the still-molten core of the Earth are classified by geologists as igneous rocks.

From the very beginning, natural forces have worn down these rocks, carrying the debris into the seas, and the sediments thus formed have been compacted and altered, forming the sedimentary rocks of the Earth's crust. There has been endless change. As land was uplifted and then eroded, sedimentary rocks were being reduced to sediments again, and new sedimentary rocks were being formed.

8 A road cutting near the new Parliament House in Canberra, showing sedimentary rock layers, folding of strata and other features which tell a geologist the history of sedimentation. *(P. R. Evans)*

8

Because the crust was originally thin, and convection currents were strong in the super-heated core, crustal movements were cataclysmic at first. They became less dramatic as the Earth cooled and the crust thickened. The addition of sedimentary rock layers, and vast accumulations of igneous material poured out by the volcanic activity which accompanies crustal movements, helped to stabilise the cooling planet. There was uplifting, downwarping, crumpling and folding of rock strata. Cracks, or faults, developed, allowing blocks of the crust to move vertically and horizontally in relation to each other.

Our Earth, as we see it now, is the product of continuous cycles of rock formation, erosion and crustal movements throughout the ages.

The layers of sedimentary rocks in the crust can be "read" by earth scientists rather like the pages of a book, and the history of their formation can be interpreted. The study and correlation of rock layers is the science of Stratigraphy.

Modern technology has enabled the dating of some of the minerals of which the

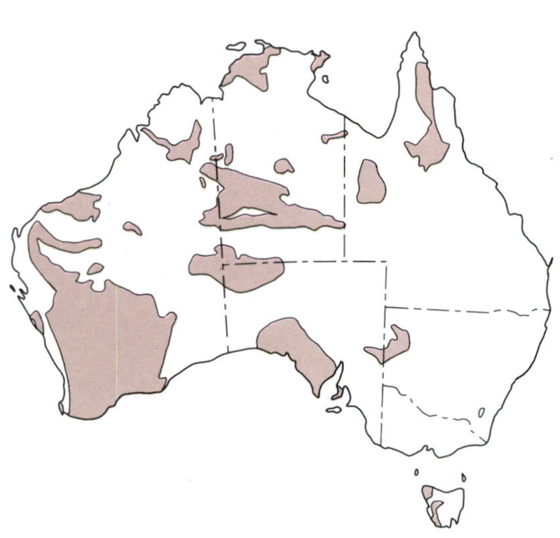

AUSTRALIA'S SKELETON OF ANCIENT ROCK UNITS ESTABLISHED BEFORE THE PHANEROZOIC.

rocks are composed, and we know that some of the oldest crystals found anywhere on Earth are in Australia. Zircon crystals which go right back to the formation of the first crust have been identified in quartzites at Mt Narryer in the Murchison region of Western Australia. They were derived from sediments between 4100 and 4200 million years old, and the rocks in which they are incorporated are themselves more than 3600 million years old. Scientists at the Australian National University in Canberra, using a new ion microprobe instrument developed in the Department of Earth Sciences, made this exciting discovery which confirms the amazing antiquity of the land surface in southern Western Australia.

Australia has a skeleton of very ancient rocks formed during the Cryptozoic Eon. Some of the very old sedimentary rocks contain the very first records of life on our planet. Although that earliest life was very different from most life as we now understand it, it set the scene for later developments.

THE DAWN OF LIFE

At the time of the dawn of life, the Earth's crust was hot and so too were the seas overlying it. Ultraviolet radiation from the sun was lethal except under deep water. The thin atmosphere was a toxic mixture mainly of hydrogen, ammonia, methane and carbon monoxide, with little or no free oxygen. Lightning was a major force in that world of volcanic upheaval, and is still associated with volcanic activity today. It would be hard to imagine conditions which would seem more hostile to life, yet it was then that life began.

9 A lightning strike. The photograph was taken on Fraser Island by the light of the strike. (Glen Carruthers)

9

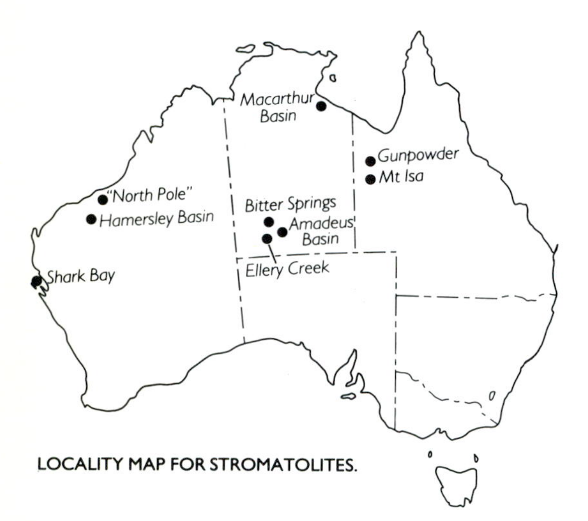

LOCALITY MAP FOR STROMATOLITES.

It has been shown experimentally in recent years that it is possible to create complex organic molecules under laboratory conditions. By passing an electrical discharge through a mixture of gases and water vapour of the same composition as the putative atmosphere of the Earth 3500 million years ago, under intense ultraviolet radiation, simple sugars and amino- and nucleic-acids have been synthesised — the building block of proteins. These experiments show how organic molecules were probably formed under natural conditions in the primordial oceans. Lightning then would have provided the electrical discharge.

The "experiments" thousands of millions of years ago were on a global scale, resulting in the formation of great concentrations of such molecules. When sufficient levels had been reached, interaction between the different molecules would have resulted. From one of these reactions came the beginning of life, with the formation of DNA (deoxyribonucleic acid, the main constituent of chromosomes, which carry the genetic code).

DNA is the fundamental building block of life, with power to replicate itself and to act as a blueprint for the formation of amino acids (the elements of proteins). The oldest life forms (and the first fossils), the Bacteria, had the same qualities, and from such microscopic beginnings all later living things have come.

THE FIRST LIVING ORGANISMS

10 A thin section of a Stromatolite seen under a microscope, showing the domes formed by the layers of lime precipitated by Cyanobacteria. (M. R. Walter)

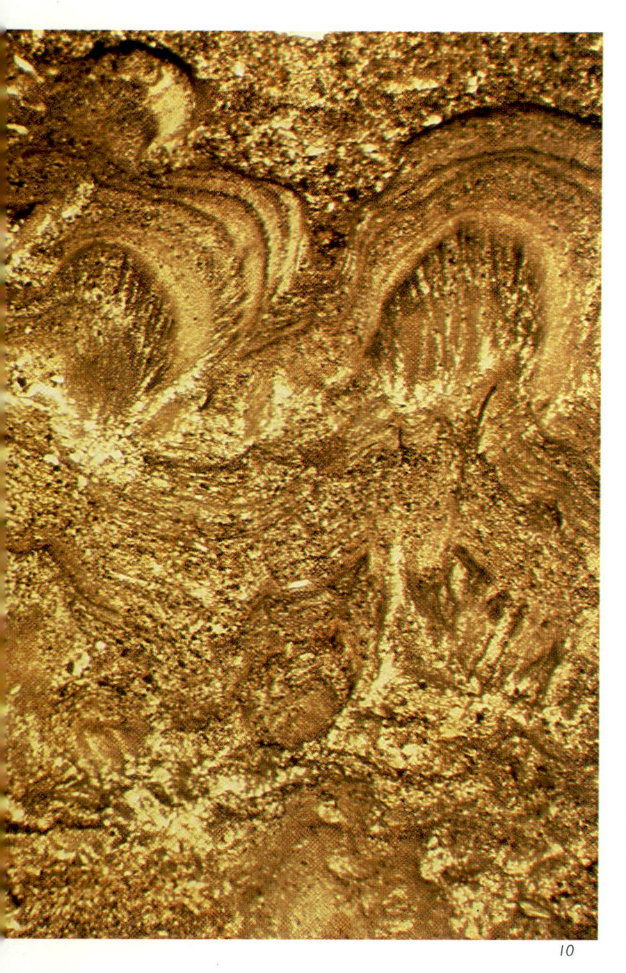
10

The first ancestral Bacteria fed on the carbon compounds which had accumulated in the seas over the millennia. They were the Heterotrophic Bacteria, and as they became more numerous their food supplies dwindled. Eventually some — the Autotrophic Bacteria — started to manufacture their own food within the cell using free hydrogen, which is produced in quantity during volcanic eruptions, or hydrogen sulphide (H_2S). They had a primitive light-harvesting pigment known as bactero-chlorophyll, which enabled them to use solar energy to combine carbon dioxide (CO_2) and hydrogen to make simple sugars. They were the first photosynthesisers, and the high levels of ultraviolet radiation in the sunlight compensated for the low efficiency of the ancestral-chlorophyll which they contained. Their activities did not alter the atmosphere in any way significant for later life forms.

Fortunately for us, and for other life today, the next stage in the evolution of living things started processes which were to change the Earth and make it ready for transformation into a "living planet".

Microscopic Bacteria-like organisms, the Cyanobacteria, which were more highly evolved than the first ancestral Bacteria, developed the ability to extract hydrogen from water while photosynthesising. (They took the H from H_2O, releasing the oxygen as a by-product.) They used solar energy to synthesise their sugars and they had a more efficient light-harvesting system than their ancestors. They had chlorophyll, which is the same green pigment used by all members of the Plant Kingdom (with the exception of some saprophytes or parasites which no longer make their own food).

As a result of chlorophyll-aided photosynthesis, oxygen was released in such quantities over such long ages that an atmosphere suitable for higher life forms was created. With rising levels of oxygen in the atmosphere, the depth of the ozone layer increased progressively and acted as a filter to screen out more of the harmful radiation, further preparing the Earth for the explosion of life which was to transform it.

Today there are still simple Cyanobacteria with the same simple structure (or lack of structure). They have no nuclei in their cells, no cell inclusions, and their green pigment (chlorophyll "a") is disseminated throughout the cell, not confined to chloroplasts. Photosynthesis occurs throughout the cell, not in special granules. They reproduce by simple cell fragmentation.

In Australia, and elsewhere in the world where ancient rocks have survived the ravages of erosion, there are fossil reefs formed by Cyanobacteria which precipitated lime and trapped sand and silt in their meshes. Also known as Stromatolites, these ancient limestones show evidence of their biological origins. Thin sections prepared for examination under a microscope show the microscopic single cell or thread form of the Cyanobacteria.

It was originally thought that the oldest reefs had been formed under deep water, which filtered out much of the harmful radiation believed to be a limiting factor for life before the ozone layer was sufficiently developed. There is now sedimentological evidence which shows that even the most ancient reefs were periodically exposed. In other words, they grew in intertidal situations. Scientists working in the field have also revised their view on the effectiveness of the ozone layer: formerly believing that it provided a sufficient filter for harmful radiation only after the Cyanobacteria had created relatively high levels of oxygen in the atmosphere, they now assume that there was some screening by an ozone layer even in earliest times and that the Cyanobacteria tolerated the high levels of radiation which were characteristic of early Earth.

The oldest Stromatolites so far dated occur in northern Western Australia at the "North Pole" locality, and are 3500 million years old. Similar aged reefs have been described from Zimbabwe, and worldwide there are many others nearly as old. Reefs dated at 1600-1700 million years old are found at Gunpowder, near Mt Isa in Queensland. [2] [3] [4]

At Shark Bay in Western Australia, Stromatolites are today being produced by living descendants of ancient reef-forming Cyanobacteria. Here, in an embayment partly cut off from the open sea, where the salinity is extremely high, weird-shaped structures occupy the intertidal zone and are exposed at low tide. It has frequently been stated that they survive, virtually as living fossils, here and in a few other rare habitats such as in the Bahamas because the salinity is too high for other life forms to survive and thus they face no competition. Indeed, the arrival of Gastropods (shelly, grazing Molluscs) and similar creatures in the Fossil Record coincides with the start of the Stromatolites' decline, and this group of Invertebrates is absent from the Stromatolites' modern habitats. However, it is now established that many other creatures have adapted to the high salinity and today live with the Stromatolites, including a number of Fish and Shrimps. [5] [6]

Living sub-tidal Stromatolites have also been discovered in Western Australia. They are never exposed to the air and they grow under conditions of low light intensity at considerable depths. Instead of being composed exclusively of mats of Cyanobacteria, as are the intertidal examples, the sub-tidal variety are compound structures of Diatoms, Green Algae and Cyanobacteria. They manage to live in equilibrium with grazing Molluscs and boring Invertebrates which share their habitat and feed on them.

11 The North Pole locality, Western Australia. (M. R. Walter)

12 Stromatolite heads separated by sandstone, in an interesting outcrop at Ellery Creek in the Amadeus Basin, Central Australia. This Jay Creek Limestone shows the flat tops of the Stromatolites as they grew, with the gaps between filled by sand. (John Webb)

13 Fossil Reef in a landscape at North Pole, Western Australia. (M. R. Walter)

EARLY UNICELLULAR LIFE

Classification of simple life forms which were characterised by the absence of a nucleus, the Prokaryotes, is at best highly speculative. Life remained at a single-cell stage for a very long time. Whilst a number of authorities believe that some cells became nucleate about 1400 million years ago, the majority believe that this

14

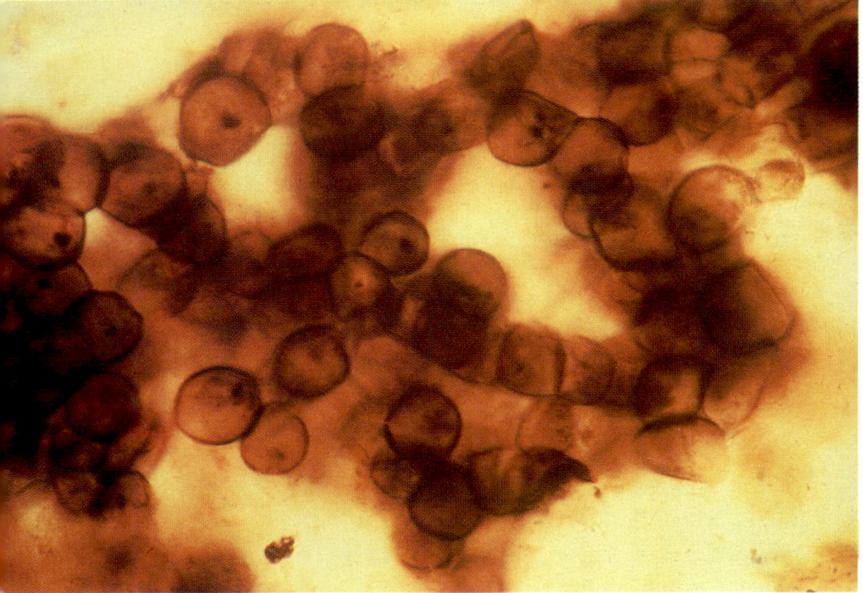

15

16

17

STROMATOLITES

The Cyanobacteria, the microscopic organisms which formed Stromatolites from at least 3500 million years ago and continue to do so right up to the present day, were the first photosynthesisers to obtain hydrogen by splitting the water molecule (H_2O) and liberating oxygen into the atmosphere. They were able to use solar energy to synthesise simple sugars from hydrogen and carbon dioxide because they had chlorophyll, the light-harvesting green pigment which characterises members of the Plant Kingdom.

Their activities created an atmosphere with sufficient oxygen to sustain the air-breathing life forms which are now characteristic of Planet Earth. As the oxygen concentration increased, the ozone layer became denser and filtered out the ultraviolet radiation which would have been higher than could be tolerated by the life forms of the Phanerozoic Eon which was to follow.

Thus it was the simple Cyanobacteria which prepared our Earth for life as we know it and determined the form it would take.

18

14 Stromatolites weathered out of the surface of limestone in the Macarthur Basin, Western Australia. They are known as Conophyton *because of the conical structures. (M. R. Walter)*

15 Cells of the organism Glenobotrydion aenigmatis *from the Bitter Springs Formation in the Amadeus Basin, Central Australia. The age of these organisms is Late Proterozoic, about 1000 million years. (M. R. Walter)*

16 A slice of North Pole reef, Western Australia, showing domed layers. Age about 3500 million years. (Jim Frazier)

17 Shark Bay Stromatolites, Western Australia. These modern Stromatolites grow as mushroom-shaped structures in the intertidal zone, in an embayment cut off from the open sea. The largest heads in the picture are 2 metres wide. The structures were formed by Cyanobacteria. (M. R. Walter)

18 Subtidal Stromatolites at Carbla, Western Australia. These examples are compound structures of Diatoms, Green Algae and lime-secreting Cyanobacteria, growing under water. (Steve Parker)

important advance did not occur until about 600 million years ago at the start of Phanerozoic time.

In modern classifications it is now accepted that there are four major "kingdoms" of living things: Kingdom Bacteria, the Plant Kingdom, the Animal Kingdom, and Kingdom Fungi.

At the simple unicellular level of organisation the distinction between the members of the kingdoms is not always clearly defined. Cyanobacteria have chlorophyll and photosynthesise, behaving as plants. Unicellular Algae are often highly mobile, resembling animals, and some engulf other organisms and digest them like animals do, as well as making their own food. The Fungi always depend on other organisms to provide their food, either as dead organic matter where the Fungi are saprophytic or by feeding via living organisms where they are parasitic. The Animal Kingdom depends on the Plant Kingdom to supply its manufactured foods at some stage in every food chain.

From earliest times there was probably some co-operation between different life forms and some interdependence. Symbiotic relationships, which result in mutual benefit, were to become of great significance between land-plants and Fungi.

Some scientists go so far as to say that symbiosis was the key to the development [7] of the more advanced plant cell.

All green plants have cells which comprise an outer membrane, a nucleus in which the genetic material is aggregated, one or more chloroplasts containing the green pigment, rod-like mitochondria (microscopic structures involved in burning oxygen to produce energy), and cytoplasm (the jelly-like material which fills the rest of the space). It is suggested that a single-celled organism engulfed a Cyanobacterium, which became its chloroplast, and that it ingested some rod-like Bacteria as well, which became its mitochondria. The three types of organism found the unit thus formed to be of mutual advantage and started to act as a single organism. The aggregation of genetic material into a nucleus came later. Some support is given to the theory by modern analytical techniques which have shown that the chemistry of the chloroplast is similar to Cyanobacteria, while that of the mitochondria resembles Bacteria.

The first nucleate plant cells were unicellular Algae. There may have been several evolutionary lines even among very early Algae, leading to the different types of Seaweeds and unicellular and other microscopic types which exist today. The type of chlorophyll in the different groups varies. Ancestral Cyanobacteria had only "a" type chlorophyll, Green Algae have "a" and "b", and the chromophyta have an additional "c" type. Very primitive Green Algae, known as Prochlorons, have "a" and "b" type chlorophyll, and have been suggested as possible ancestors for Green Algae. It is possible that the chlorophyll types may show evolutionary lines in plants as being genetically controlled in the same way as blood groups show relationships in humans, but at this stage the whole area is highly speculative.

The Green Algae are central to this account of the origins of the Australian land-flora, as they are believed to be ancestral to all Higher Plants.

THE AGE OF EVIDENT LIFE

ERAS OF THE PHANEROZOIC EON

The Phanerozoic Eon is divided into three Eras: the Palaeozoic (ancient), the Mesozoic (medium age) and the Cainozoic (young). These Eras are further sub-divided into Periods which are given different names as shown on the geological Time Column. The enormous amount of evidence and information which has been accumulated by earth scientists over the last two hundred years has enabled accurate dating of these geological Periods.

The activities of the earliest primitive life forms had altered the environment, providing an atmosphere for the oxygen-breathing organisms which are characteristic of life on Planet Earth. The ozone layer which resulted, filtered out much of the harmful radiation from the sunlight, setting the stage for the explosion of life that was to come.

The history of the Living Earth is told by rocks formed during the Phanerozoic Eon, which represents the most recent 12 per cent of the geological Time Column. Some of the sedimentary rock layers formed during this Eon contain evidence of plant and animal life in the form of fossils. By studying the fossils that occur in the different layers of stratigraphical sequences, palaeobotanists and palaeontologists have been able to determine much about the stages of evolution.

Various life forms in the evolutionary sequence of both plants and animals arose

GEOLOGICAL PERIODS OF THE PHANEROZOIC EON

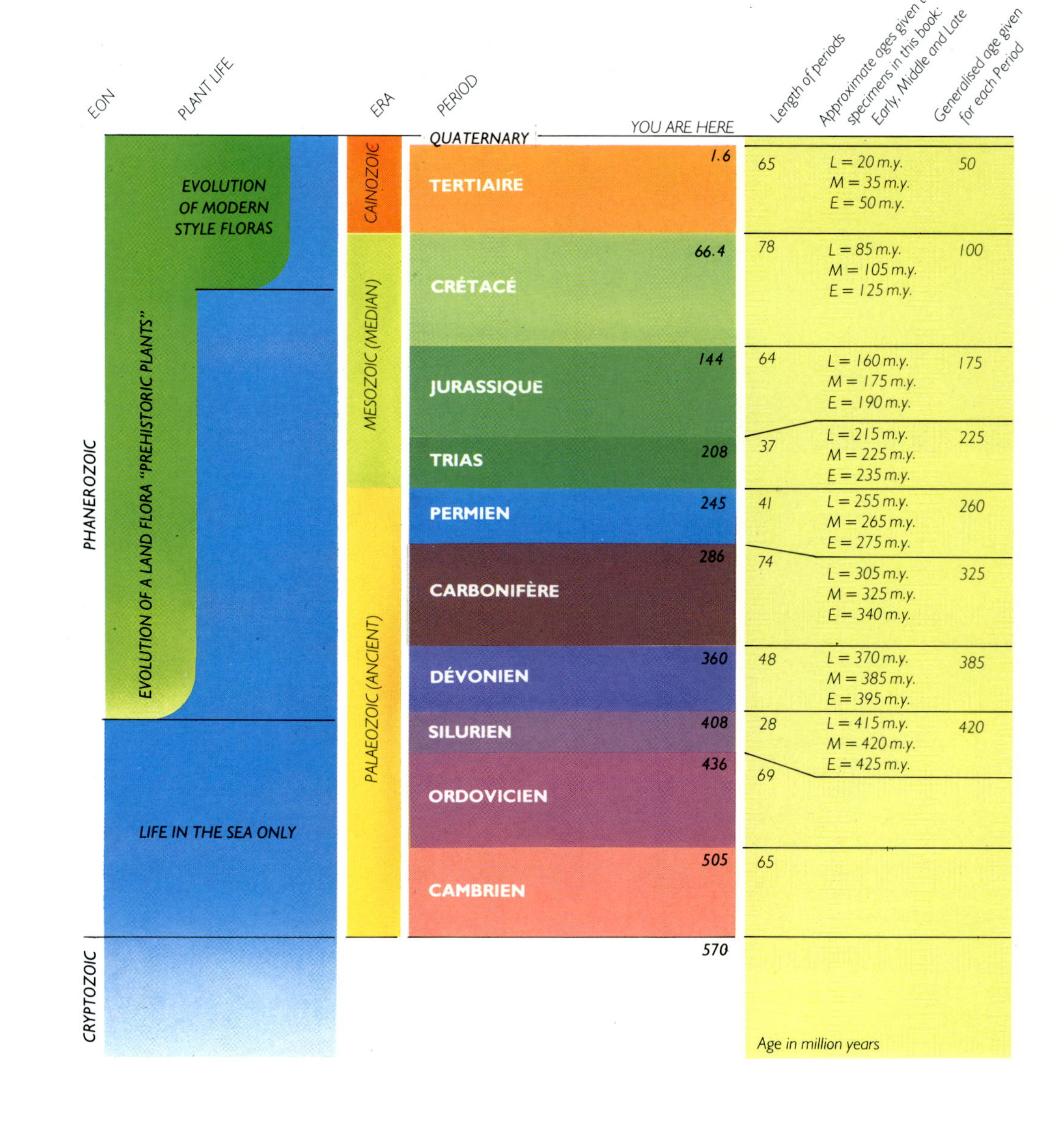

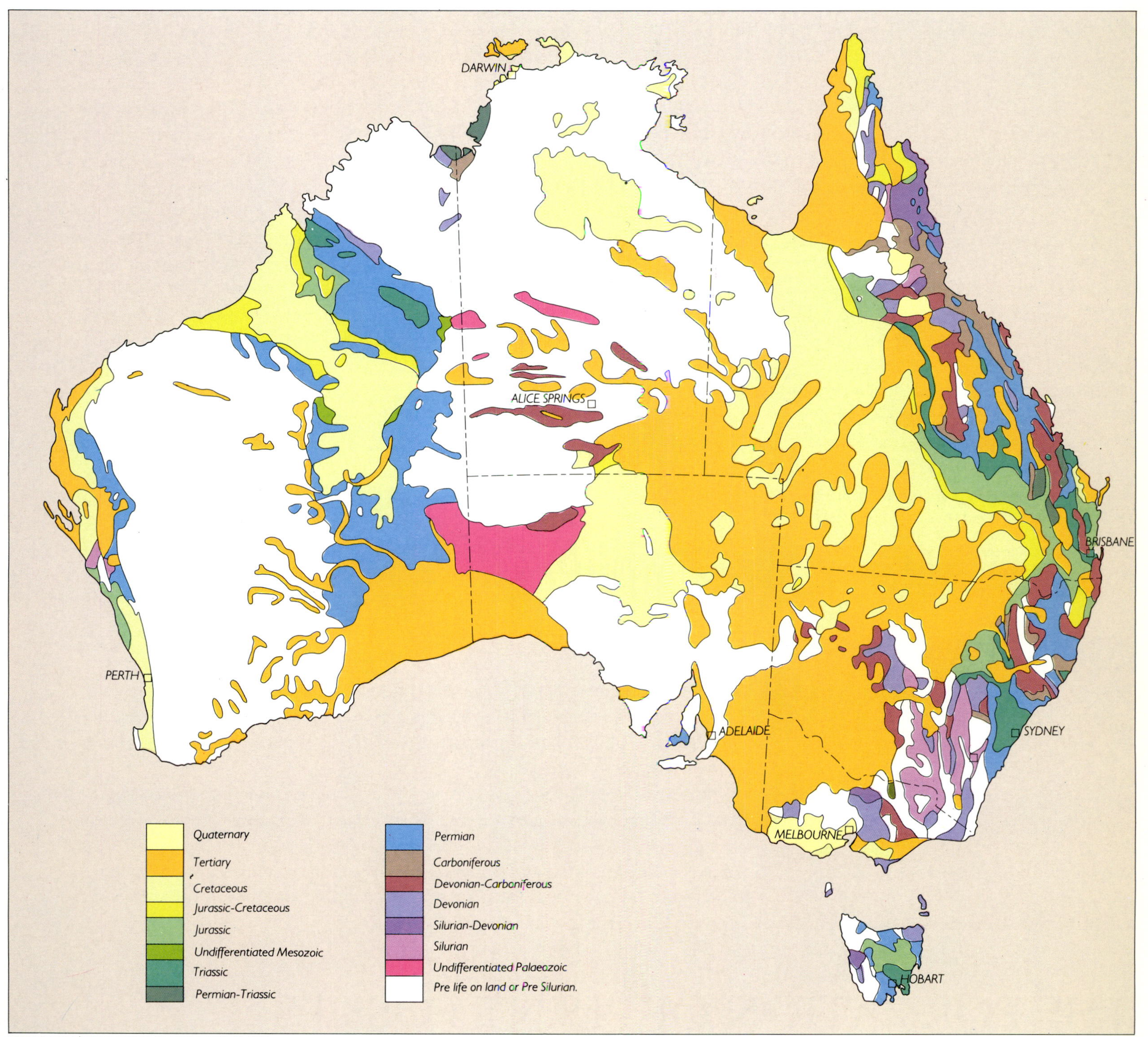

SIMPLIFIED GEOLOGICAL MAP OF AUSTRALIA.

at well-defined times, flourished and then became extinct. They were followed by other forms which also had limited time ranges. It is possible, therefore, to correlate rocks by identifying the fossils they contain; the presence of a particular fossil in a specific rock layer tells us the age of the sediments which formed that layer.

Sedimentary rock sequences were laid down in sedimentary basins — downwarps in the Earth's crust into which rivers drain, carrying sediments eroded from rocks of the local land surfaces. Under suitable conditions, carbonates (which form limestones) may accumulate in the basins as a result of the biological activities of aquatic organisms in the waters of the basin itself. The accumulating weight of sediments causes further sagging of the crust and layers (or strata) of sediment are compacted and converted into a sedimentary rock sequence.

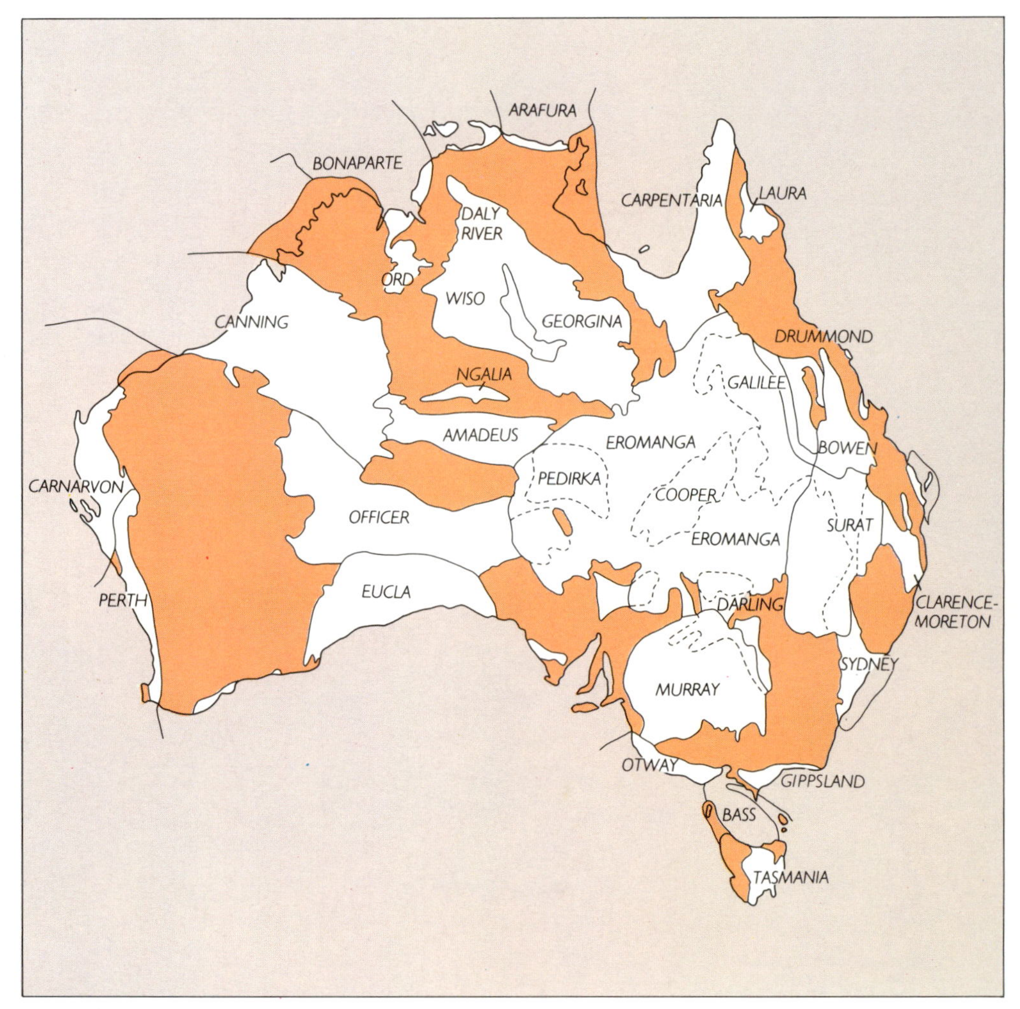

PHANEROZOIC SEDIMENTARY BASINS.

19 A microscope slide section showing the cell structure of petrified wood of Dadoxylon farleyense, a Permian tree trunk from Farley, New South Wales. A central, disorganised pith cavity and an annual ring of radially arranged wood elements can be clearly seen, but outer tissues have been disturbed in preparation of the microscope slide. Approximately 260 million years old. (× 3.4)

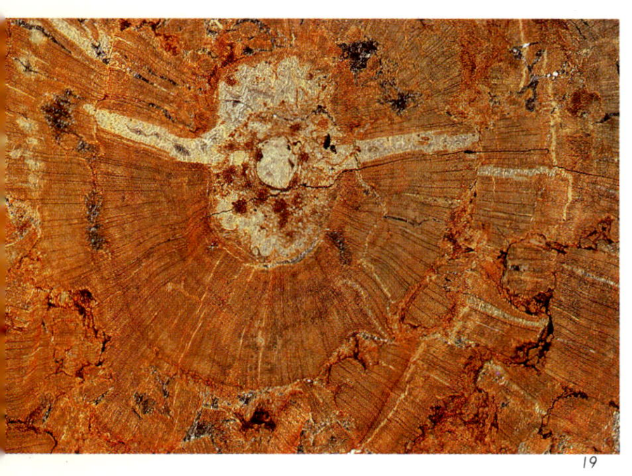
19

Some sedimentary basins owe their origin to faulting and rifting of the crust during tectonic movements. During this massive cracking and tearing, whole blocks may subside while others are elevated, and drainage into the lowered regions allows sediments to gather.

Most plant fossils collected in Australia are from sedimentary basin sites. Many of them have been found through systematic mapping of areas in the course of exploration for coal, oil and gas. These fossil fuels owe their presence to organic materials incorporated in sediments when the rocks were being formed, and Australia is unique in that the plant material from which its oil and gas reserves are derived is of terrestrial and not marine origin.

THE FORMATION OF PLANT FOSSILS

There are two types of fossils: those that are visible to the naked eye, known as macrofossils; and those which can be seen only under a microscope, known as microfossils.

MACROFOSSILS — PETRIFACTIONS, IMPRESSIONS AND CASTS

When sediments from eroding rocks were washed into basins and accumulated there, to be transformed by compaction and pressure into rock layers, plant debris was often carried with them. Leaves, bits of stem, pieces of wood, cones, seeds, roots or even whole small plants or tree trunks were rapidly covered by sediment and the processes of decay were prevented or at least retarded. In the course of time,

20

21

20 A slice of a small, petrified tree trunk in which cracks and cavities have been filled with precious opal. This specimen from the opal fields at Lightning Ridge in New South Wales is part of the Galman Collection in the Australian Museum. It is of Cretaceous age, approximately 100 million years old.

21 Part of a petrified tree trunk, the top of which had been cut and polished. The diameter of the trunk is about 35 cm. Perfect preservation of bark and wood makes this fossil look like a living tree trunk. Replacement of plant tissue by mineral matter has resulted in replication of minute cellular details. From the Cretaceous of the New England District of New South Wales, age approximately 100 million years.

if conditions were favourable, some of the plant material might become fossilised. High levels of acidity, such as occur in peat and other swamps, aid preservation.

Most plant fossils occur in freshwater deposits or shallow marine basins.

The coal deposits of the world are accumulations of enormous volumes of vegetation that grew in swamps, some of which persisted in the same area for many millions of years. The weight of the accumulating material depressed the hollows in which it was collecting, and in some cases great depths of coal seams were formed. However, while deep burial and great pressure have converted the carbon into coal, details of the plants which supplied the carbon have been largely obliterated.

As a rule it is only in the layers of shales and mudstones associated with coal deposits that well-preserved plant macrofossils occur. In some cases, though, mineral-rich water has percolated through swamp deposits during the process of coalification, causing some of the plant material to become petrified. The molecule-by-molecule replacement of plant matter by minerals results in perfect preservation, in which every microscopic detail of a plant is faithfully reproduced.

Petrifactions can occur under a range of conditions, not just in coal swamps. Most examples are pieces of "petrified wood", but there are also "petrified forests" in which tree trunks and their roots have been preserved *in situ* following some cataclysmic event, like burial in volcanic ash or hot mud after a volcanic eruption or rapid covering by sediments in a great flood. Some petrified wood is preserved as agate and rare examples as opal. Although most petrified wood does not attain gem quality, polished sections can be very beautiful.

Thin sections can be cut from petrifactions so that the detailed structure and arrangement of cells can be examined under a microscope.

The more delicate parts of plants are very rarely seen as petrifactions. Consequently, much of our knowledge of plants of the past is superficial, being obtained from other kinds of fossils in which microscopic detail is usually not preserved.

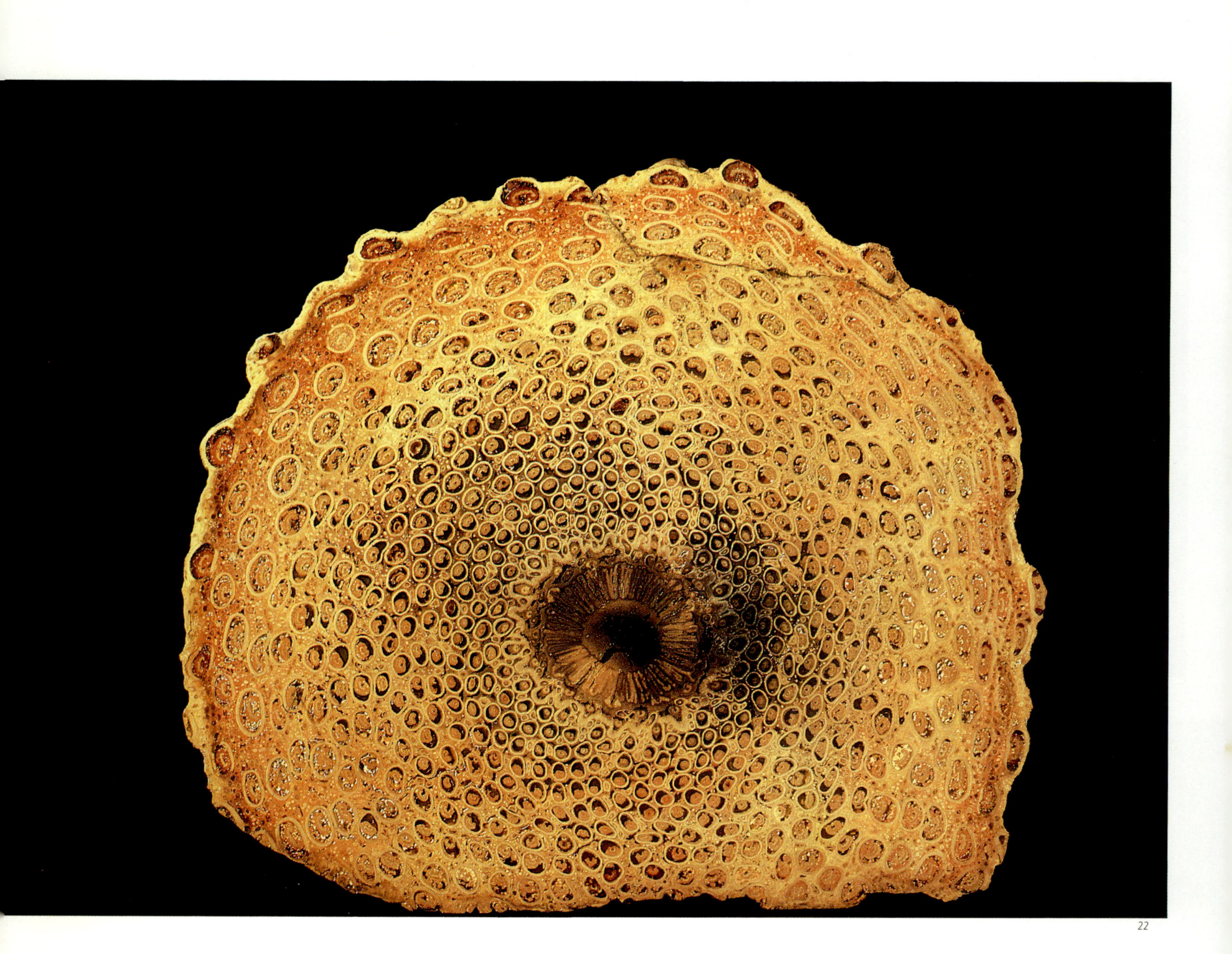

22

22 Palaeosmunda sp., a petrified Tree-fern trunk seen in cross-section. The "trunk" is a compound structure of leaf bases, roots and main stem, as in modern Tree-ferns. From Lune River in Tasmania, age probably Mesozoic.

23 A cut and polished wedge of Tree-fern "trunk", Palaeosmunda sp., from Lune River, Tasmania. The core of vascular tissue is seen at the top of the specimen. The pattern of rings in the outer tissues is made by the vascular traces which supply the fronds, and the smaller spots between them are roots in cross-section. This specimen is 6 cm long and has a maximum width of 3 cm (seen here greatly enlarged).

For instance, when a leaf or other plant fragment is entombed in sediment, it is flattened during the compaction process. Usually all the original plant material disappears and only its imprint is left on the layer of sediment. This imprint is an "impression", and most leaf fossils are of this sort. A carbon film may remain on the impression surface, or sometimes the cuticle (waxy protective skin of the leaf) may be preserved.

Cuticles can be removed from the rock, prepared and examined under a microscope. They supply useful information on the epidermal cells and the pores (or stomata) through which gaseous exchanges take place, and can even tell us about the relationships of the plant which bore the leaf. Some cuticle characters are specific to particular plant groups.

If the plant fragment undergoing fossilisation were bulky and three-dimensional, it would probably undergo considerable flattening and deformation. Soft objects would be particularly vulnerable. The type of fossil which results from a bulky object can show a variety of forms. A stem may be preserved as a cylinder with surface markings intact and may sometimes be freed from the rock when it is split open, leaving hollow impressions. Such a fossil is a "cast". A variant of this form of fossil is an "internal cast", in which sediment in the hollow core of a stem shows the form of the inside cavity. If a cast is firmly embedded in the rock so that only one side is

exposed when the specimen is split, then the half cast has a counterpart which is an impression.

Mostly it is the harder, more resistant plant fragments which are preserved as fossils. Delicate parts are less likely to survive transportation into sediments and more likely to decay.

Only the finer-grained sediments in a rock sequence are likely to contain well preserved fossils; shales, mudstones and very fine-grained sandstones are the usual rock types. The finer the grain, the more minute detail is visible. Also, the more readily the rock splits into layers (as shale does), the easier it is to expose the fossils.

The study of plant fossils is complicated by the fact that only rarely does one find a complete plant, or find a connection between different organs that establishes without doubt which parts went together to form a particular plant. In the naming of plant fossils, generic and specific names are given to the different organs because they are almost invariably found separately. In most cases it is only the ubiquitous association of parts — leaves with stems, or stems with roots, or fruits with leaves, etc — which leads to an educated guess that they were parts of one sort of plant. For instance, it is now known that *Glossopteris* leaves, *Araucarioxylon* wood, *Vertebraria* roots, and *Austroglossa* and *Squamella* reproductive structures are parts of the Glossopterid trees which were the dominant plants of Permian times.

The gradual accumulation of knowledge has resulted in a body of information which now enables us to reconstruct many of the ancient plants with a fair degree of confidence. However, there is still much which has to be found out before we can have a proper understanding of our floral heritage, and there are gaps which will probably never be filled.

The amount of information that fossils can supply is limited by their type of preservation as well as by their fragmentary nature. Furthermore, the Fossil Record is patchy. The chances of a plant species leaving a fossil record at all, unless it existed for a very long time, are slight. Only those plants which grew in habitats close to water would be likely to appear in the record at all. And those that grew in habitats where the fossils were actually being formed would be disproportionately represented, giving an incorrect impression of the composition of the vegetation of the times.

24

24 An impression fossil of a twig of Agathis jurassica, a Kauri Pine, and a Fish, Leptolepis talbragarensis, from the Jurassic at Talbragar, New South Wales. Age approximately 175 million years. No organic material is preserved. The fossil is only an imprint on the surface of the rock layer, and silica has been deposited on the impression, making it white.

25 Three small, opalised cones from Lightning Ridge, New South Wales, in the Galman Collection at the Australian Museum. The largest has a maximum width of 2 cm. The two larger cones may be Araucarian, but the affinities of the small, narrow one are unknown. Age Cretaceous, about 100 million years.

26 Cast of a stem. A stem of a Giant Clubmoss, Leptophloeum australe, from the Late Devonian at Barraba, New South Wales. Age about 370 million years. The rhombic leaf-base pattern on the stem surface is characteristic of this species. (× 1.5)

27 Cuticle, the waxy layer on the outside of the epidermal layer of a leaf. Occasionally cuticle may be preserved as a film on an impression surface. It can be separated from the rock, treated to remove mineral matter, bleached and stained, and mounted on a glass slide for examination under a microscope. Cuticles give detailed information on the arrangement of epidermal cells, the nature of stomata and the arrangement of veins. The small, dark spots between veins on this Dicroidium cuticle from the Triassic (about 225 million years ago) are glandular hairs.

25
26
27

MICROSCOPIC PLANT FOSSILS — SPORES AND POLLEN

28 A cloud of pollen being blown from male Pine cones. When we see the number of pollen grains produced by one cone we can appreciate how aerial plankton can contain a spore sample from a whole countryside. Pollution monitors make spore counts in air samples and warn asthma sufferers when pollen counts are high, as in spring when many plants are releasing their clouds of pollen. Wind-blown pollen was incorporated in sediments in past ages, making the Pollen Microfloral Record. Prior to the evolution of the Flowering Plants, with their pollination often carried out by members of the Animal Kingdom, wind was the almost universal agent of distribution.
(Glen Carruthers)

While macrofossils form the visible and often beautiful record of plants through the ages, there is also a microscopic record which supplies a great deal of information about vegetations of the past. Called the Spore and Pollen Record, it is a much more useful tool in stratigraphy than are the macrofossil assemblages.

Spores and pollen grains have waxy, cuticularised outer skins (or exines) which are decay-resistant. Plants produce vast quantities of pollen or spores, as anyone who suffers from hay-fever will testify. Clouds of these almost invisible grains form aerial plankton which is carried in air currents and blown by the wind, circulating freely, often over immense distances. (Today, clouds of pollen from Australian She-oaks are blown across the Tasman Sea to Mt Cook in New Zealand, where they colour patches of ice on the glaciers.)

When sediments accumulated in the past and formed sedimentary rocks, they incorporated a sample of the spore and pollen population of the time. Because of the wide dispersion, a much more representative selection from the vegetation is shown than is the case with macrofossils.

The study of fossil spores and pollen is the science of Palynology. Palynologists crush samples of sedimentary rocks and extract the spore material. Some sedimentary rock layers have rich microscopic floras, some are barren, and in some the spores have been damaged or destroyed by weathering or other processes to which the rocks have been subjected. Good samples can produce so many spores that statistical analyses can be made.

Seen under a microscope, spores and pollen grains are often surprisingly complex and beautiful. The sculpturing and ornamentation of the exines is different in the different plant species which produced them. As fingerprints can identify an individual person, so a spore can identify a species. But the problem is not as simply solved as in the human situation. Although a spore can be given a specific name and identified wherever it occurs in samples of different rocks, for the most part it cannot be assigned to a specific fossil plant. Nevertheless, spores of the distinct plant groups and families have characteristic features, and we know that certain types were produced by plants which had leaves belonging to certain palaeobotanical genera.

When palynologists are dealing with the pollen of the Flowering Plants in the upper parts of the Time Column (such pollen appears in the record only from 100 million years ago) it is often possible to equate fossil pollen with that of living plants. A complete picture of a flora with this modern aspect can therefore be obtained. It is only by understanding the fossil Pollen Record that one can obtain an accurate picture of the history of the modern Australian flora.

An assortment of Australian fossil pollen grains of the Mesozoic and Cainozoic eras. (× 800)

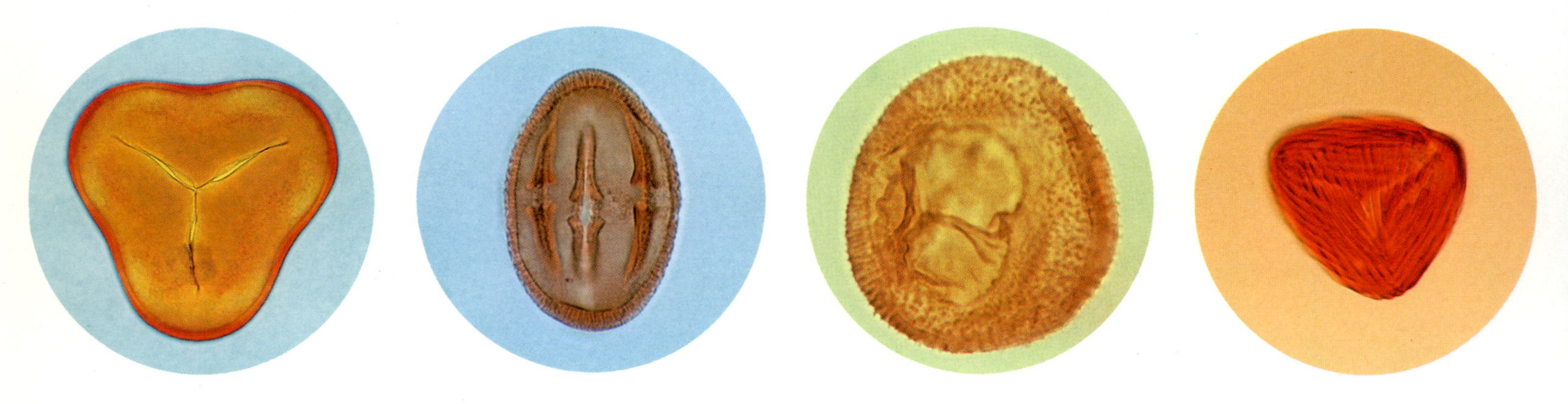

28

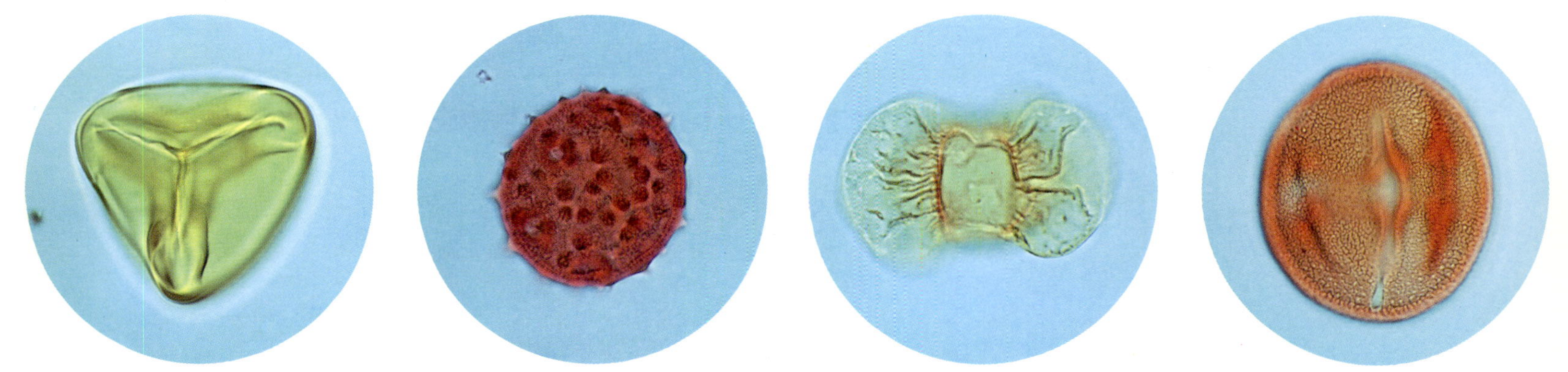

AUSTRALIA IN GONDWANA

29

30

Today the continents of the Southern Hemisphere are widely separated. Early European explorers had expected to find *Terra Australis*, a large continental mass surrounding the South Pole in the manner in which the landmasses of the Northern Hemisphere are aggregated around the the North Pole. Instead they found vast oceans and isolated lands — for they were more than 100 million years too late. The "Great Southern Continent" of their mythologies did not exist. But scientists were soon to realise that during geological time, configurations of land and sea had changed.

It would seem that the observations of early explorer-scientists, who noted the similarities of vegetation in isolated southern lands, started the modern theorising about moving continents and former supercontinents, and gave birth to a science which has proved that Australia was part of a great southern landmass for most of Phanerozoic Earth history.

The similarity of the circum-Antarctic floras of Kerguelen, Tierra del Fuego, the Falkland Islands, south-eastern mainland Australia, Tasmania and New Zealand was difficult to explain in terms of long-distance dispersal mechanisms. Disjunct distributions were initially said to have come about because of former land-bridge and island-chain connections between the now widely dispersed lands.

Joseph Hooker, naturalist to James Ross's Antarctic voyages in the *Erebus* and *Terror* during the years 1839-1843, was the first to notice the essential unity of the modern circum-Antarctic flora. He discussed his theories of a former "Great Southern Continent", which included a vegetated Antarctica, with Charles Darwin. The implications which he saw — that the existing distribution of plants is dependent on former dispositions of land and sea areas — was revolutionary for its time.

Hooker influenced Darwin's thinking, as is seen in the chapter on geographical distribution in the *Origin Of Species* where Darwin wrote: "I am inclined to look in the Southern, as in the Northern Hemisphere, to a former and warmer period, before the commencement of the last glacial period, when the Antarctic lands, now covered in ice, supported a highly peculiar and isolated flora." There was no fossil evidence at that time to indicate how long ago the "former" period might have been.

Plant fossils, however, had been collected in Antarctica prior to Hooker's voyages. James Eights, aboard the *Seraph* during the United States Exploring Expedition of 1838-42, under Charles Wilkes, recorded the presence of fossil wood in the South Shetland Islands but the discovery did not impress the scientific community of the day.

8

Scientific interest in an Antarctic flora was not to be re-awakened until the end of the century, and then only following a new awareness of an Arctic flora in the Northern Hemisphere.

This "Northern Arctic Flora" had been described by Oswald Heer in 1868. Considerable interest in Heer's views was aroused when Captain Larsen of the Norwegian barque *Jason* in 1894 recorded, from Antarctica, "frequent" specimens of petrified wood with annual rings "which looked as if they were of deciduous trees". It was predicted that a similar temperate flora to that of the Arctic might be found at the South Pole because of the world climatic conditions implied by the northern flora.

A few years later the Swedish South Polar Expedition of 1901-1903 collected Jurassic plants from Hope Bay. These specimens showed clear similarities with the

31

29 *Microscope section of petrified wood of a Glossopterid tree trunk, seen in cross-section, from Beaver Lake, Antarctica. Of Permian age, about 260 million years. A number of annual rings show that growth patterns were seasonal. The wood is composed of very regular tracheids. (× 4.0)*

30 *Longitudinal section of the wood from Beaver Lake, Antarctica. (× 40)*

31 *Antarctic landscape. (Alex Ritchie)*

fossil assemblages already known from Jurassic strata in Europe. Their discovery profoundly influenced scientific thought in regard to past climates and distribution of ancient life forms.

Around the same time "traces of plant life" were collected by the geologist T. Ferrar on Captain Robert Scott's first Antarctic Expedition in 1901-1904, but they were dismissed as insignificant by the palaeobotanists Arber and Seward, in 1907 and 1914 respectively. However, in 1928 these specimens were re-examined by W.N. Edwards, who in splitting them exposed new plant remains which he was able to identify as *Glossopteris* species of Permian age (i.e. about 260 million years old).

By this stage, though, *Glossopteris* plants had already been firmly established in the history books following Scott's ill-fated second South Polar Expedition of 1911-12. Edward Wilson, physician and scientist to the expedition, collected 35 lbs (16 kg) of rock specimens from "coal seams containing leaves and stems" in cliffs at Mt Buckley, at the head of the Beardmore Glacier, on the return journey from the Pole in February 1912. When the bodies of Scott, Wilson and Bowers were discovered in their tent by the rescue party the specimens were found with them. The rocks were sent to England, where they were reported on by Seward. He recognised both the palaeoclimatic implications of the extension of Gondwana to South Polar regions and the significance of the occurrence of *Glossopteris* in relation to the former union of Antarctica and Australia.

THE THEORIES OF CONTINENTAL DRIFT AND PLATE TECTONICS

GONDWANA, THE GREAT SOUTHERN CONTINENT.

With the discovery of similar fossil floras and faunas in India and all the southern continents, particularly Antarctica, the theory of Continental Drift was [9] postulated and the concept of a supercontinent comprising South America, Africa, [10] Madagascar, India, Antarctica, Australia and New Zealand came into being. The name Gondwana, which is applied to this supercontinent, means "land of the Gonds" and is derived from a region of the same name in India where the *Glossopteris* flora was first described. Earlier the term "Gondwanaland" was generally used, but it is tautological and at the 1980 Fifth International Gondwana Symposium in Wellington, New Zealand, delegates officially resolved to adopt "Gondwana" instead.

The Continental Drift theory neatly explains the fossil similarities, the geological evidence and the distribution of some modern plants and animals that exists in the southern lands. Furthermore, the re-assembly of the continents into one landmass showed a good fit. But the mechanisms by which movements of bits of the Earth's crust could have occurred could not be explained, so for some time the theory did not gain universal acceptance. However, great advances in technology, especially in connection with studies of palaeomagnetism, established the new concept of Plate Tectonics.

The theory of Plate Tectonics shows that the Earth's crust comprises at least fifteen "plates" on which the continents ride. The plates are rigid and virtually undistorted slabs of lithosphere, and the major ones occupy considerable areas of the globe.

The boundaries of the plates are of four main types: divergent or spreading zones, where plates are separating and new plate material is being added; subduction zones, where plates converge and the edge of one plate is forced under and consumed; collision zones, which are former subduction zones where the continents riding on the plates are colliding; and transform faults, where two plates are gliding past one another with no addition or destruction of plate material.

DISTRIBUTION OF FAUNA AND FLORA WHICH CONFIRM THE CONCEPT OF GONDWANA.

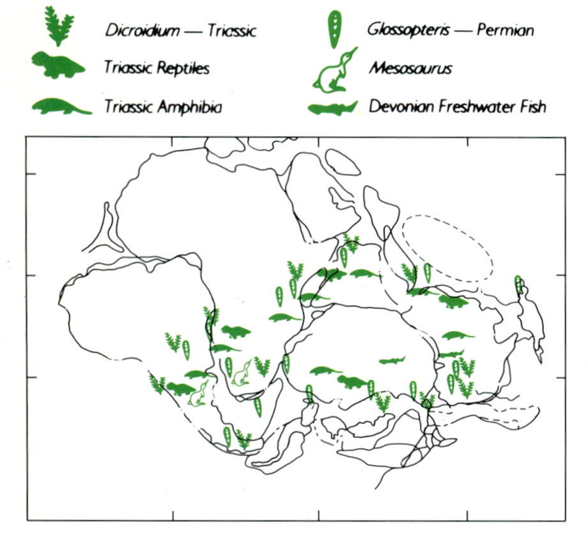

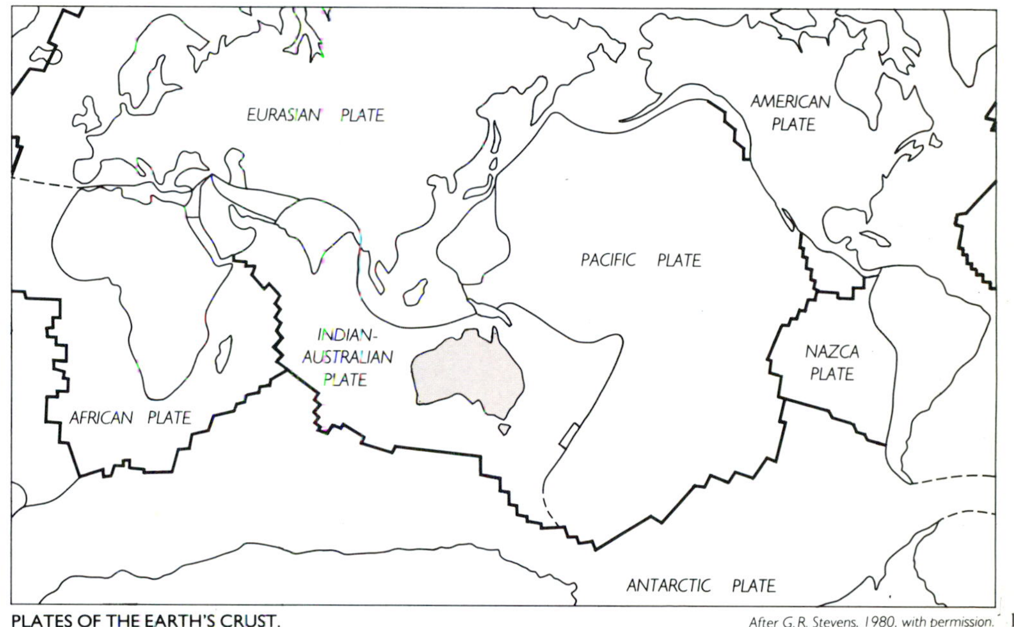

PLATES OF THE EARTH'S CRUST. *After G.R. Stevens, 1980, with permission.* 106

32 Gangamopteris cyclopteroides, a Glossopterid from South Victorialand, Antarctica. This Permian plant, approximately 260 million years old, occurs in southern Africa, India and Australia, as well as in Antarctica. The Glossopterids, with their Gondwanan distribution, are part of the main fossil evidence which confirms the concept of the Great Southern Continent. (× 1.5)

32

Active zones of the Earth's crust — where earthquakes, volcanic activity and mountain-building are occurring — lie on the boundaries of plates and are related to the movements between them. The active volcanoes of today are primarily on the edges of plates. Volcanoes that occur in the middle of a plate, such as in Hawaii, are related to deep-seated "hot spots" in the mantle.

With the theory of Plate Tectonics clearly explaining how the plates of the Earth's crust have moved, altering the configurations of land and sea over the ages, there is now a growing interest in the theory that the Earth itself is expanding during the process. According to the "Expanding Earth" proponents, 180 million years ago, in the Jurassic Period before the Gondwanan split-up occurred, the diameter of the Earth was 80 per cent of its present measurement. By 90 million years ago, in the Cretaceous, its diameter was 90 per cent of its present size; by 60 million years ago, in the Palaeocene, it was 93 per cent; and by 30 million years ago it was 97 per cent.

The understanding of the fundamental mechanics of crustal movements, which have resulted in the rearranging of the continents through time, has firmly established that the "Continental Drifters" were right. And, not only had there been a "Great Southern Continent", or supercontinent, but it was preceded by an even larger supercontinent known as Pangaea in which the southern landmasses of Gondwana were joined to the northern supercontinent of Laurasia.

For most of its long Fossil Record history, Australia was part of the supercontinent Gondwana, which according to the reassembly used in this book included the major landmasses of South America, Africa, Antarctica, India, Australia-New Guinea and New Zealand. (It also incorporated Madagascar, Arabia, Turkey, Iran and Afghanistan. Parts of China and South-East Asia probably formed part of the supercontinent before they moved off and joined Asia.)

Australia's contact with other lands remained until relatively recent times, and not until the last links were severed about 50 million years ago did this land become the island-continent. From then onwards, Australia become increasingly isolated as it moved north away from Antarctica. The northward movement is calculated to be about 7 centimetres per year. In the last 45 million years the continent has moved through 27° of latitude. In 20 million years from now the tip of Cape York will be on the Equator and Hobart will be on the latitude of Sydney today.

33 *Glossopteris sp.*, from South Victorialand, Antarctica. The Glossopterids are associated with coal in Antarctica and other Gondwanan lands. (× 3.7)

34 *Dicroidium sp.*, from Argentina. The *Dicroidium* flora occurs in all the southern lands which were previously part of Gondwana. Age Triassic, about 225 million years. (× 3.0)

35 *Dicroidium sp.*, from South Africa, which was also part of the Triassic Gondwanan province. (× 4.0)

36 *Glossopteris indica*, a Glossopterid leaf from India. The same species occurs in all the continents which are fragments of Gondwana. Age Permian, about 260 million years. (× 1.3)

37 *Dicroidium odontopteroides* from the Horseshoe Mountains, Antarctica. The *Dicroidium* flora flourished throughout Gondwana during the Triassic, when the climate was warm to hot (even in Antarctica) and there were probably times of aridity. (× 1.8)

38 *Vertebraria indica*, a root of a Glossopterid plant. These roots with segmented internal structure are characteristic of this Order of gymnospermous plants. The segmentation is believed to have been related to an aerating function as the Glossopterids grew in swamps. *Vertebraria* occurs wherever Glossopterid leaves are found. Age Permian, about 260 million years, from Antarctica. (× 2.7)

33

36

34

37

35

38

AUSTRALIA'S GONDWANAN HERITAGE

39 *Rainforest, Russell River, North Queensland. (Glen Carruthers)*

40 *A Strangler Fig which starts to grow high up on a forest giant and sends down its roots in a curtain around its host. It uses this strategy to get up into the light — competition for light and space being intense in such a densely populated ecosystem. Epiphytes such as Orchids, Ferns and Aroids grow on its roots, all striving for more light and better conditions than they would find on the forest floor. (Densey Clyne)*

As Australia was part of other landmasses until relatively recent geological times, most of the country's fossil history is cosmopolitan or Gondwanan. It is only since the southern continents started to separate and move into their present positions that the fossil flora and fauna became Australian. This separation of continents took place at the time when there was a changeover underway from ancient-style floras to modern-style vegetation, consequently present-day distribution of the Flowering Plants reflects the changes which were occurring in the distribution of land and sea in the last 100 million years.

Worldwide, modern floras are dominated by the Flowering Plants, or Angiosperms. The ancient cosmopolitan communities of the Mesozoic Era — Conifers, Cycads, Ferns and other groups with long fossil records — were largely ousted during Early Tertiary times. The Angiosperms appeared suddenly on the scene, evolved and diversified rapidly, and were capable of filling most niches so successfully that they changed the whole aspect of world floras.

Nearly a hundred years ago Charles Darwin described the origin of the Angiosperms as an "abominable mystery", and much of the mystery still remains — though theories abound.

It is generally accepted that western Gondwana was probably the region where the first Angiosperms arose. At that time, not more than 120 million years ago, the North African-northern South American region of the supercontinent was increasingly affected by aridity. In climatically transitional areas on the margins of the warm and wet ancient forests, where the increasing seasonality of rainfall pushed the boundaries of the existing vegetation type back, evolutionary pressures were strong and the first Flowering Plants came into being.

The progenitors of modern-style world floras were able to radiate out into the whole of Gondwana to the south and also spread north into Laurasia. The origin of the ancestral Angiosperms (with their great potential for evolution and adaptation) at a time when the big break-up and movement of continental landmasses was getting underway, explains some modern plant distribution patterns.

Common ancestry in a relatively small number of original Angiosperm lines explains the relationships of different floras at the Order and family level in both Hemispheres. Later, separate and parallel evolution from the initial stock explains the distinct characteristics of individual floras.

It was during the early radiation of ancestral Angiosperms that Australia in Gondwana received its quota of primitive Flowering Plants. A sub-tropical migration route existed across India before that country moved off north about 90 million years ago, breaking the route. A temperate route through South America and Antarctica remained open for about another 40 million years. From the early Angiosperms, which came in by both routes, the Tertiary Australian Rainforest floras developed, and after the changeover to a vegetation dominated by Flowering Plants, Rainforest clothed much of the continent for more than half of the Tertiary Period.

Australia's most obvious Gondwanan inheritance today can be seen in a few localised areas of relict Rainforest, survivors from Tertiary antecedents and now

confined to very limited niches in which suitable conditions prevail. Within the tropical region of the north, the tropical and sub-tropical regions of the north-east and the eastern temperate region are small "islands" of Rainforest (or Closed Forest), of several different types.

Closed Forest vegetation, now restricted to this "archipelago of refugia", as Dr Len Webb describes the relictual habitats, was comparatively widespread from 60 to 20 million years ago, then fluctuated over considerable areas during the climatic changes which have occurred up to the present.

When white man came to Australia two hundred years ago, Rainforest occupied only about 1 per cent of the total area of the continent. In the brief time since then, this vegetation type has been reduced to a quarter of its former extent and what is now left is mainly confined to areas that have been considered unsuited to cultivation or too inaccessible to be commercially exploited for timber.

The Rainforest is almost completely unrelated floristically to the typical "Australian vegetation" surrounding its well-defined habitats. It is a little-changed part of the country's Gondwanan inheritance and shows strong affinities with the Closed Forests of other countries. The component plants of the Rainforest are mostly conservative in the evolutionary sense, and include a surprising wealth of old, primitive Flowering Plants.

In general, when climate and habitat conditions do not change greatly or suddenly over time, an established vegetation type can occupy the same geographical area almost indefinitely. There is no great pressure on species to evolve further. If gradual climatic changes occur, making much of the area unsuitable for the original ecosystem, communities become isolated in refuge areas where local conditions are suitable. From these refuges they can spread back into wider areas if conditions change and become favourable again.

"Refugia" may be a disjunct chain of islands of vegetation, as is the case with Rainforests today. Larger scale changes might involve alterations to the boundaries of vegetation zones in response to changes in global weather patterns. Margins of deserts advance and retreat, grasslands replace shrublands and scrub replaces forest, or vice versa. On a smaller scale, vegetation is always a shifting mosaic of smaller units with specialised ecological requirements within the framework of vegetation zones.

Climatically transitional zones have been especially important as centres for plant evolution. Here the pressures to evolve new strategies to exploit the new set of conditions is strong, and change is the order of the day.

41 *"Cauliflory", a condition where flowers are borne on the trunks of trees, another adaptation to Closed Forest living. (Densey Clyne)*

42 *Only the rivers break the canopy in the Closed Forests, giving a glimpse of towering trees, stratified layers and epiphytes. (Densey Clyne)*

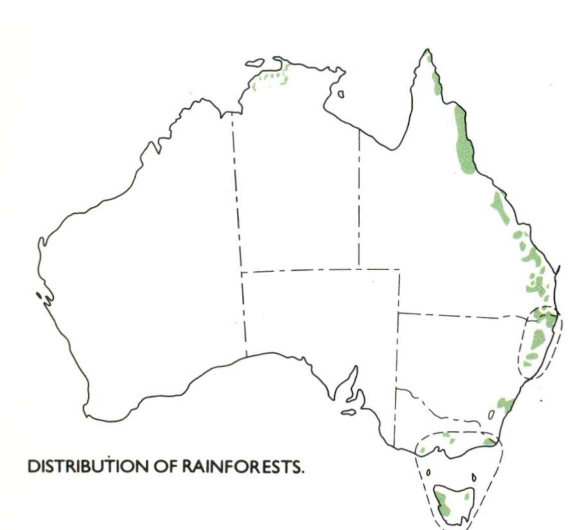

DISTRIBUTION OF RAINFORESTS.

EVOLUTION OF AN AUSTRALIAN FLORA

From the Gondwanan stock which had become established in Australia by the time separation from the rest of Gondwana was complete, the modern Australian flora has developed in isolation for about 30 million years without any significant input from migrants. Only in the last few million years has there been contamination, particularly in the north by plants invading from neighbouring countries.

The characteristically "Australian" vegetation is derived from the ancestral Gondwanan flora. Eucalypts in dense or open forest, in dry woodland or scrubland; Acacias growing as Wattle trees in woodland or as Mulga in arid regions; Casuarinas on river banks; hummock grasses, or "spinifex", on the wide, dry plains of the Centre and north; Grass-trees and Banksias in the heathlands: all are totally Australian and as distinctive as the Kangaroos and other Marsupials which populate the land.

The special character and uniqueness of the Australian flora is due mainly to the omnipresence of the genus *Eucalyptus*. No other comparable area of land in the world is so completely characterised by a single genus of trees as Australia is by its gum-trees. Acacias are almost as widespread and visible.

Apart from the two main dominants there are genera and species which are endemic (they do not occur naturally anywhere else). In fact, 80 per cent of all the plant species and 30 per cent of the genera occurring in Australia today are endemic.

43 *Grass-trees are among the strange Australian endemics, having evolved here in the time when the continent was isolated from other lands after separation from Antarctica. They are adapted to life on poor soils and to fire, and regenerate after bushfires. When purchased from a nursery for planting in gardens, a blow-torch is often used to char the outside of their trunks to stimulate growth and make them thrive in their new situations. (Densey Clyne)*

43

The famous European botanist Joseph Hooker, writing about the Australian flora in his introduction to *A Flora of Tasmania* in 1860, stated: "It contains more genera and species peculiar to its own areas, and fewer plants belonging to other parts of the world, than any other country of equal extent. About two-fifths of its genera and upwards of seven-eighths of its species, are entirely confined to Australia."

This high rate of endemism accounts for the distinctive Australian character of the flora. When seen in a world context, however, the Australian plants are clearly related to other world floras. The families of plants represented in this country occur widely elsewhere. Relationships of Australian flora to that of the southern continents of Africa and South America, and to New Zealand and other isolated southern islands, are clearly evident.

12

The particular and distinctive quality of the Australian flora is the adaptation resulting in leathery, hard, spiny or reduced leaves. This adaptation is known as scleromorphy and the plants whose leaves are adapted in these ways are called sclerophylls. Apart from the very limited broad-leaved Rainforests and the wet sclerophyll forests of the high country of south-eastern Australia and southern Western Australia, the vegetation is of dry sclerophyll type.

The generally accepted view among botanists is that scleromorphic adaptations originated in plants evolving on the margins of the Closed Forests as the climate became drier and the rainfall more seasonal. Rainforests are more or less closed ecological systems, creating their own microclimate, enriching their own soils, and remaining in balance as an entity. Because of the rate of decay of vegetable matter in a Rainforest, the nutrient supply is constantly recycled and an equilibrium exists. Evolutionary pressures in such a stable community are small.

44

MAP OF PRINCIPAL FLORISTIC ZONES WHICH CORRESPOND TO CLIMATIC ZONES AND RAINFALL PATTERNS.

14

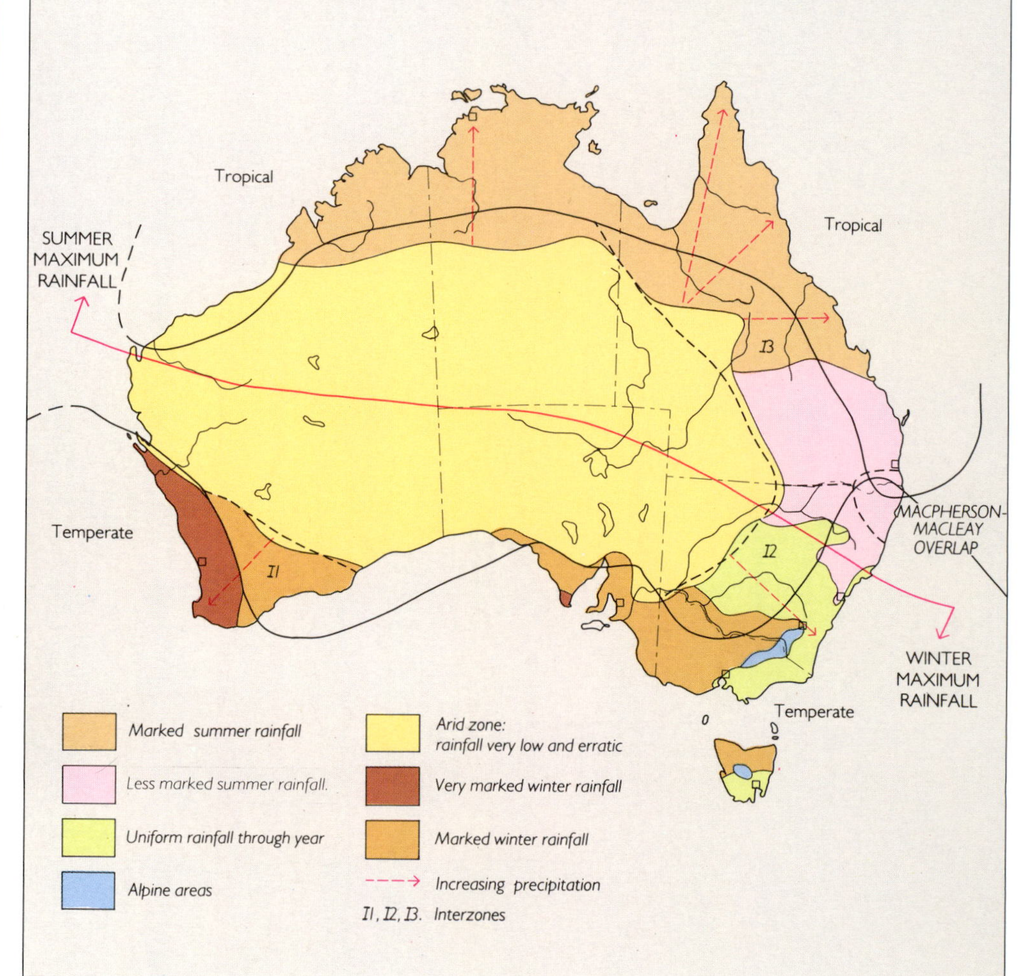

45

44 *Wet Sclerophyll forest. Giant Eucalypts characterise this forest type, whereas relict Rainforest are broad-leafed forests without Eucalypts. (Densey Clyne)*

45 *Hakea sp., family* PROTEACEAE, *with small, spiky leaves, woody stems and dry, bristly flower head. (Densey Clyne)*

46

48

47

46 *Mallee, a shrubby Eucalypt with leathery leaves, brilliant red, fleshy fruits. The stems arise from underground lignotubers which are protected from fire and drought, allowing the plant to regenerate after its aerial parts have been destroyed. (Densey Clyne)*

47 *Red-flowered Eucalypt. The long, narrow, leathery leaves are designed to hang at an angle to the sunlight so that as little of the surface faces direct light as possible. Stems are woody, the flower bud is sealed by a cap which falls when the flower opens, and the gumnuts are woody. (Densey Clyne)*

48 *Gum trees shed little shade because of the shape of their leaves and the way they hold them. (Densey Clyne)*

There is evidence of a climatically transitional zone acting to promote evolutionary change. On the fringes of the self-sustaining Rainforest ecosystem, there was need to adapt. Here the changing climate which was causing the contraction of the Closed Forest would be increasingly felt. Another important factor had also to be taken into account: poor soil. The ancient soils which are characteristic of this continent are nutrient-deficient after millennia of leaching without renewal from volcanic activity.

The sclerophyll adaptation is believed to be directly related to the impoverished soils, whose effects were felt immediately outside the Closed Forest ecosystem. The sclerophyll vegetation as it now exists is highly evolved and living in harmony with the "poor" soils, whose "deficiencies" are subjectively judged by European agricultural standards.

13

The Rainforests contained representatives of the plant families to which the sclerophyllous plants belong. The MYRTACEAE have many Rainforest genera. *Eucalyptus*, which belongs to that family, is not a constituent of true Rainforest but was a later derivative. There are also primitive genera of PROTEACEAE, from which the Banksias, Spider Flowers and other sclerophyllous genera have evolved. Other Rainforest families also have sclerophyll descendants.

The sclerophyll adaptations were secondarily arid-adaptations suited to the increasing dryness of the climate. In the Mediterranean-style climate of southern Western Australia there has been great evolution, diversification and radiation of specialised sclerophyllous plants — very largely endemic — in the last 5 million years, since that type of climate was established.

The Arid Zone floras take many of their species from adjacent dry sclerophyll regions, but many of the ephemerals may have had a cosmopolitan coastal origin and are less characteristically Australian.

As the sclerophyll vegetation evolved, adapting to the soils and the drier climate with its seasonality of rainfall — and producing species suited to all the different habitats with local special conditions — a new ingredient was introduced into the picture. This component was fire.

49

50

49 *Dry grassland, with Acacia trees and scrub, near Oodnadatta, South Australia. (Glen Carruthers)*

50 *Desert Oak, Casuarina with hummock grasses and Saltbush, in the arid Centre. All are plants capable of withstanding heat and drought because all exhibit sclerophylly. (Densey Clyne)*

51 *A bushfire burning unchecked in a remote and inaccessible area. (Jim Frazier)*

52 *Xanthorrhoea Grass-trees and a termite mound, in a Dry Sclerophyll community with Eucalypts. Cape York, North Queensland. (Densey Clyne)*

53 *Riverine forest with Paperbarks and Tea-trees, Myrtaceous plants adapted for living where the water table is high. On the Mary River, Northern Territory. (Bill White)*

51

52

53

In a Closed Forest vegetation, lightning strikes have limited effect. The evergreen foliage with high moisture content does not flare, and there is no accumulation of litter on the forest floors to sustain or spread a fire. In dry sclerophyll lands, on the [15] other hand, not only do the plants burn readily but there is a dry litter of leaves, bark [16] and woody twigs on the ground providing ready fuel. Because of the ubiquitous presence of Eucalypts in the open forests and scrublands, an extra dimension is present in the wild fires: the very high flammability of Eucalypts, which have volatile oils in their leaves, results in great heat intensity when they burn. The volatiles rising above the vegetation, in the heat in front of an advancing line of fire, can literally explode and fires can leap ahead over firebreaks.

Australians are constantly and often brutally reminded of the part that fire plays in the country's landscapes. Before the arrival of Aboriginal people about 30,000 years ago, the fires which moulded the vegetation were caused by lightning and other natural phenomena. With the coming of the hunter-gatherers, the incidence of fires increased. Since the arrival of white man on the scene, its effects are increasingly devastating.

Selection of fire-tolerant adaptations has always been a feature of our sclerophyll vegetation. In Eocene times, 50 million years ago, there were Banksias with fruits [17] like modern ones, having woody seeds that require fire to germinate.

Adaptations to fire are many and different. Some trees have thick, insulating bark to protect living tissues from overheating. Some use the opposite strategy and shed outer layers of the bark so that there is no build-up which can smoulder and burn after a flash fire. Special buds are produced below the bark in many trees so that when the leafy crown is destroyed by fire, new growth can start all along the trunk. Some shrubs produce big woody underground "lignotubers" which are capable of withstanding the hottest fires and sending up new branches after the ashes have cooled. Many plants produce hard, woody fruits and seeds which can not only survive intense heat but may even need the stimulus of fire to split open and start to grow. Many seeds are designed to lie for years in the soil waiting for the nutrients from ashes after a fire to start them growing. This way they germinate at a time when there is space for them, as the fire has reduced the competition.

And so, from the original Gondwanan stock, the unique modern Australian flora has evolved — adapted to the impoverished soils, to the arid regimes, and to fire.

54 Charred *Banksia* heads with their woody fruits opened like gaping mouths by the heat of the fire, enabling them to shed their seeds. Not only do they survive fires, but fire has become a necessary factor for their success in reproduction. Many members of PROTEACEAE are dependent on fire to split their fruits and let the seeds out. *(Densey Clyne)*

55 Very thick, corky bark on the outside of this *Banksia* trunk acts as insulation and protects living tissues from the heat of fires. *(Densey Clyne)*

56 A "*Banksia*-Man" fruiting head still smouldering after a bushfire. The open mouths of the fruit show seeds ready to emerge. *(Jim Frazier)*

57 After the fire, a charred landscape. Only sticks remain, and deep ash carpets the ground. There is no sign of life. *(Densey Clyne)*

58 New buds sprouting at the base of an *Angophora* trunk. Shrubby growth results, and when coupled with the production of new branches and leaves from buds which develop beneath the bark on the main trunk, the tree which regenerates has a different form and appearance from the original. *(Densey Clyne)*

59, 60 Grass-trees, or "Blackboys", regenerating. Long, thin leaves elongate from the growing regions within the trunk, carrying the charred remains of the burnt leaves as tips on the new. *(Densey Clyne)*

54

55

57

58

59

56

60

THE LINNAEAN SYSTEM OF CLASSIFICATION OF LIVING ORGANISMS

Plants are classified according to the system instituted by the eminent Swedish scientist, Linnaeus in 1753.

In Linnaeus's "binomial" system, each recognisably different plant has a generic and a specific name. Closely related plants belong to the same **Genus,** and genera are grouped into **Families** which have well defined characteristics. Families which are known to be related because of the nature of their reproductive structures are grouped into **Orders.** These in turn are grouped into **Classes,** and the Classes into **Divisions.**

With living plants, the classification is based on detailed information, and the generic and specific names are applied to complete plants. In the case of fossil plants, evidence of the complete plant is often absent so "generic" and "specific" names are given individually to the known parts of the plant — leaves, stems, roots, cones and seeds. The classification, therefore, is not strictly comparable with that of living plants, and one fossil plant may well comprise several "generic" and "specific" names depending on how many of its organs have been found and described.

Modern classification of the Plant Kingdom recognises fifteen Divisions but only five of them are relevant to the study of land floras, fossil and living. They are:

Division BRYOPHYTA the Mosses and Liverworts.

Division PSILOPHYTA the first land-plants, very important in the Fossil Record but insignificant today.

Division LYCOPHYTA the Lycopods (Clubmosses), very important in the Fossil Record but insignificant today, being represented by a few genera of Moss-like plants and Quillworts.

Division SPHENOPHYTA the Calamites and Horsetails, very important in the Fossil Record but today represented only by a few species of Horsetails.

Division PTEROPHYTA includes the three main Classes to which most modern plants belong. They are: **Class Filicinae,** the Ferns, which has a long fossil history; **Class Gymnospermae,** the Gymnosperms, which includes Conifers, Cycads and Ginkgos as well as several extinct Orders; and **Class Angiospermae,** the Angiosperms or Flowering Plants, relatively recent in the Fossil Record and now dominating world floras.

CLASSIFICATION OF THE PLANT KINGDOM

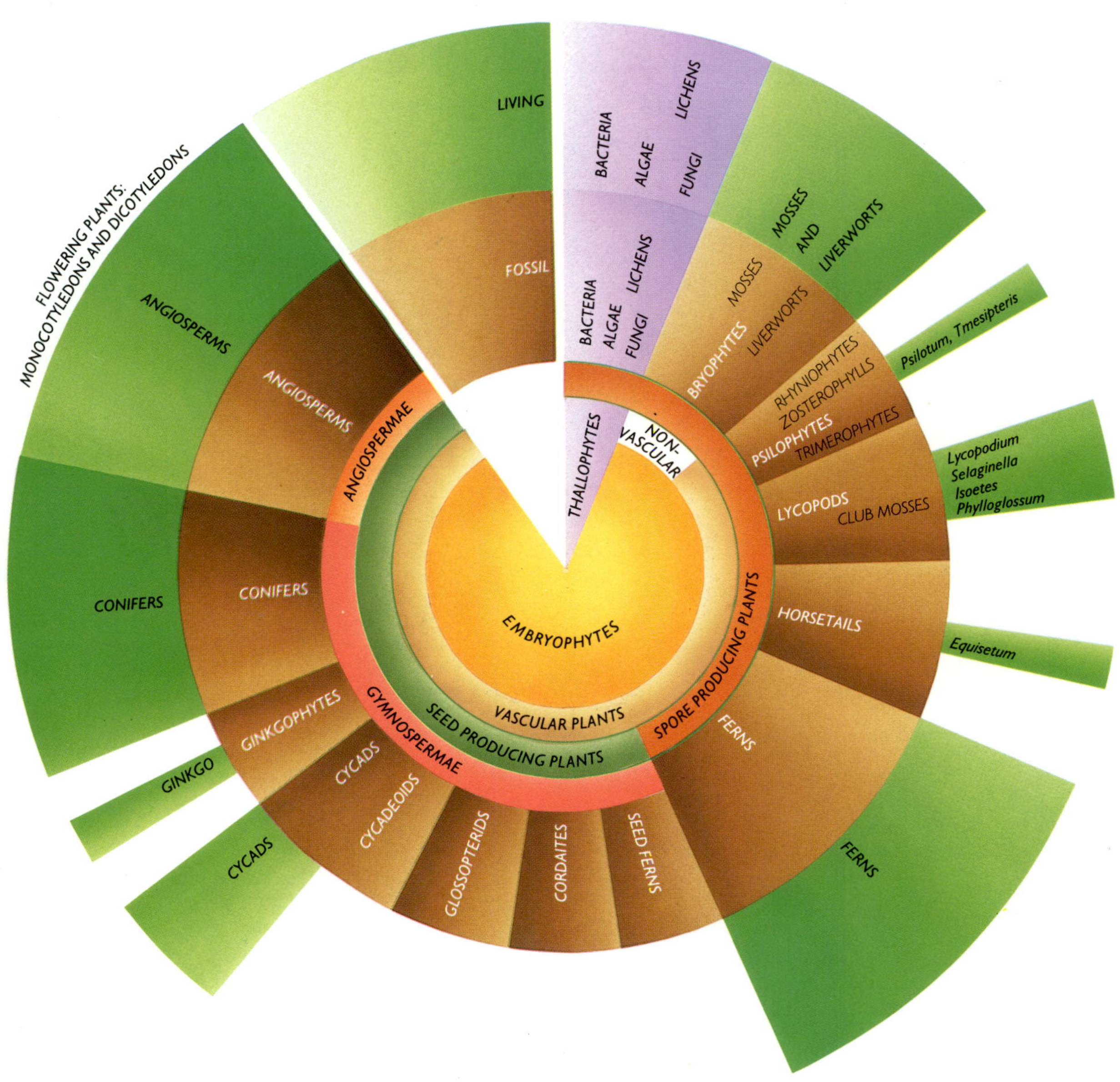

EVOLUTION OF THE PLANT KINGDOM

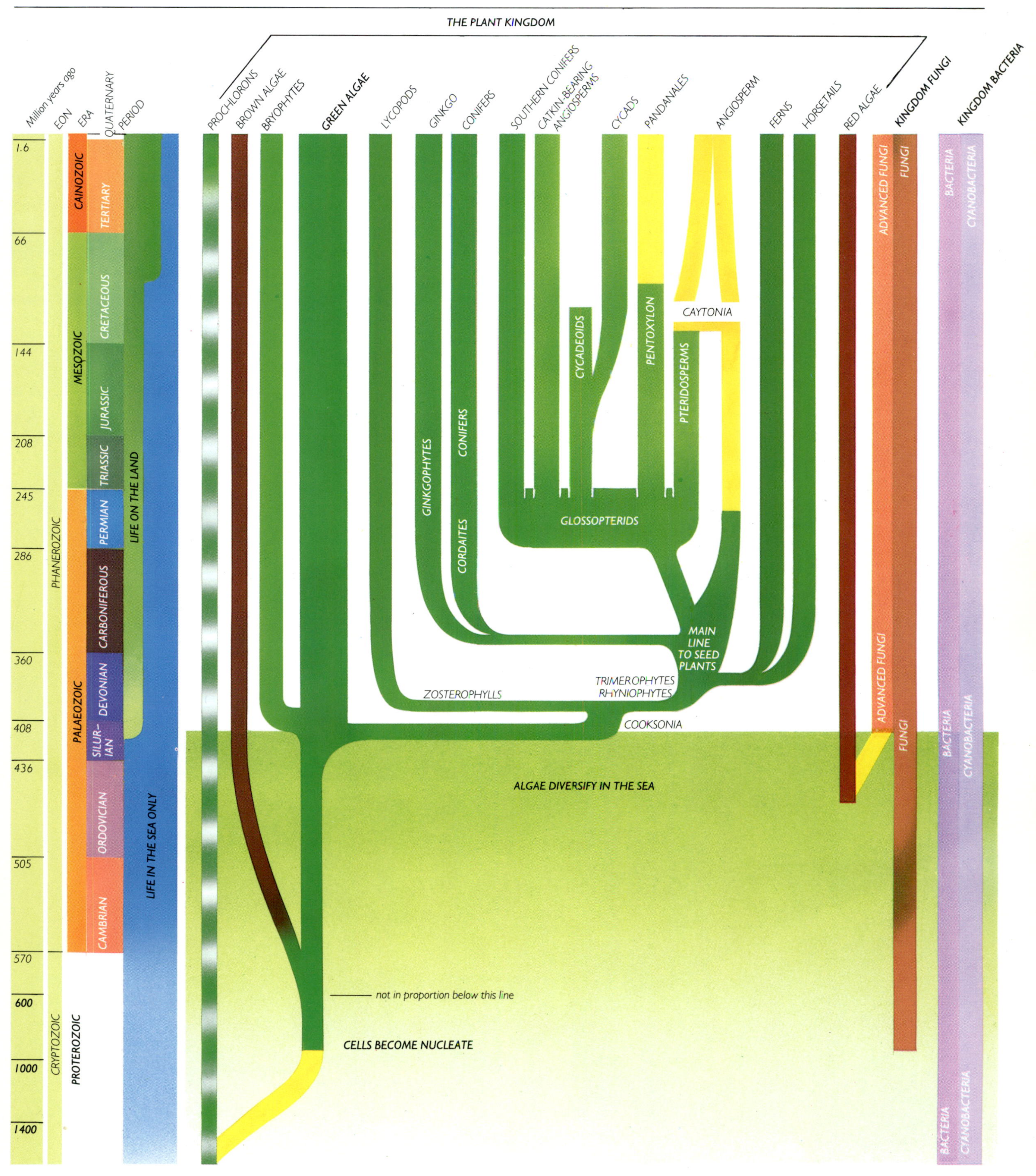

THE PLANT KINGDOM
Million years ago
EON
ERA
PERIOD
QUATERNARY
PROCHLORONS
BROWN ALGAE
BRYOPHYTES
GREEN ALGAE
LYCOPODS
GINKGO
CONIFERS
SOUTHERN CONIFERS
CATKIN-BEARING ANGIOSPERMS
CYCADS
PANDANALES
ANGIOSPERM
FERNS
HORSETAILS
RED ALGAE
KINGDOM FUNGI
KINGDOM BACTERIA
1.6
66
144
208
245
286
360
408
436
505
570
600
1000
1400
PHANEROZOIC
CRYPTOZOIC
CAINOZOIC
MESOZOIC
PALAEOZOIC
PROTEROZOIC
TERTIARY
CRETACEOUS
JURASSIC
TRIASSIC
PERMIAN
CARBONIFEROUS
DEVONIAN
SILUR-IAN
ORDOVICIAN
CAMBRIAN
LIFE ON THE LAND
LIFE IN THE SEA ONLY
CAYTONIA
CYCADEOIDS
PENTOXYLON
PTERIDOSPERMS
GINKGOPHYTES
CONIFERS
CORDAITES
GLOSSOPTERIDS
MAIN LINE TO SEED PLANTS
TRIMEROPHYTES
RHYNIOPHYTES
ZOSTEROPHYLLS
COOKSONIA
ADVANCED FUNGI
FUNGI
BACTERIA
CYANOBACTERIA
ALGAE DIVERSIFY IN THE SEA
not in proportion below this line
CELLS BECOME NUCLEATE

PART TWO

AUSTRALIA'S FOSSIL FLORA

Imagine a world in which the land is dead . . . elemental Earth and Air and Fire and Water. No signs of life. Stark beauty, awe-inspiring. Blue sea, white-horse flecked, with breakers rolling in on empty beaches, smashing on the rocks, fretting away the edges of the land. Eternal seascapes back into the mists of time, for tides have ebbed and flowed since the waters of the first oceans in the beginning felt the pull of the Moon. Great storms have whipped the waves into wild frenzy of flying spume and spindrift. And calms have followed storms, blue days and wavelets lapping on the shores . . . .

The land is rock and sand and boulder — Moon landscapes of plains and hills and ragged mountains. Volcanoes belch their glowing fire and shed their ash.

The colours are earth colours. Browns and reds and ochre, sulphur-gold and grey and black. Rain falls and wears away the naked land. The run-off rivers feed silt-clouded lakes or tumble to the sea. They carry grit and stones and mud, and pour them out into the waters.

Below the moving, windswept surface of the sea, a living world . . . an eat-or-be-eaten, catch-me-if-you-can, restless universe of creatures large and small. They live, and eat, and breed and die. Their world is water, and Seaweeds are their forests and their gardens. Their dominion ends at the margins of the land. Here, in the intertidal zone, rock pools display their microcosms, shelly creatures, Sponges and Seaweeds encrust the fringe, and almost bridge the gap between the worlds of water and of land.

But still the land is lifeless . . . .

And now the transformation . . . the first greening of the world, a tide of green which creeps and spreads across the swamps, traces the water margins, encircles the lakes and lagoons. A bloom of green which soon becomes a pelt. A spiky fur of green; a mossy carpet increasingly diverse; a vegetation which holds the skin of soil and clothes the barren rocks . . . .

And as it spreads it tempers the harshness of the ancient land, gives shade and shelter, food and opportunity, offering an invitation to animals to leave the water and follow it into the new world . . . .

CHAPTER 1

LIFE IN THE EARLY SEAS

THE CAMBRIAN AND ORDOVICIAN PERIODS TO EARLY AND MID SILURIAN TIMES

APPROXIMATELY 600 TO 420 MILLION YEARS AGO

During the first 200 million years of the Phanerozoic Eon — from the Cambrian Period through to Mid Silurian times — life was confined to the seas. The Algae were evolving and diversifying from the first unicellular organisms to multicellular plants with increased specialisation, and microscopic primitive Fungi were also present.

PREVIOUS PAGE
61 Modern Mosses and Lichens. Representatives of two very ancient groups of land-plants, which show little difference in structure and behaviour from their ancestors of 350 million years ago. (Jim Frazier)

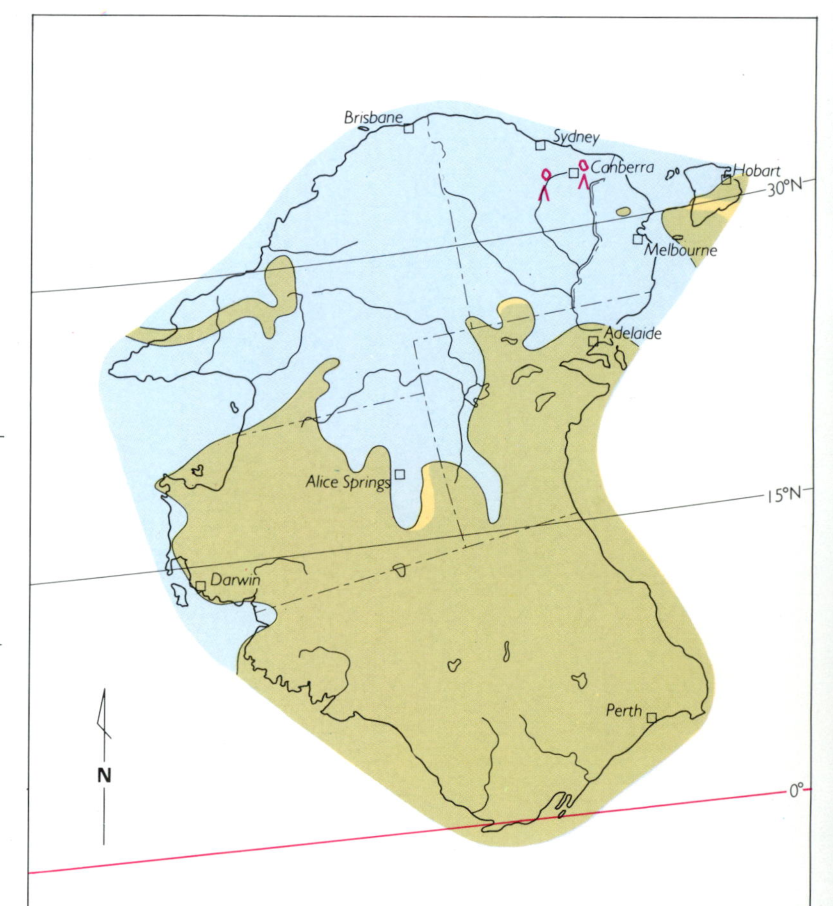

PALAEOGEOGRAPHY OF THE LATE CAMBRIAN.

PALAEOGEOGRAPHY OF THE MIDDLE ORDOVICIAN.

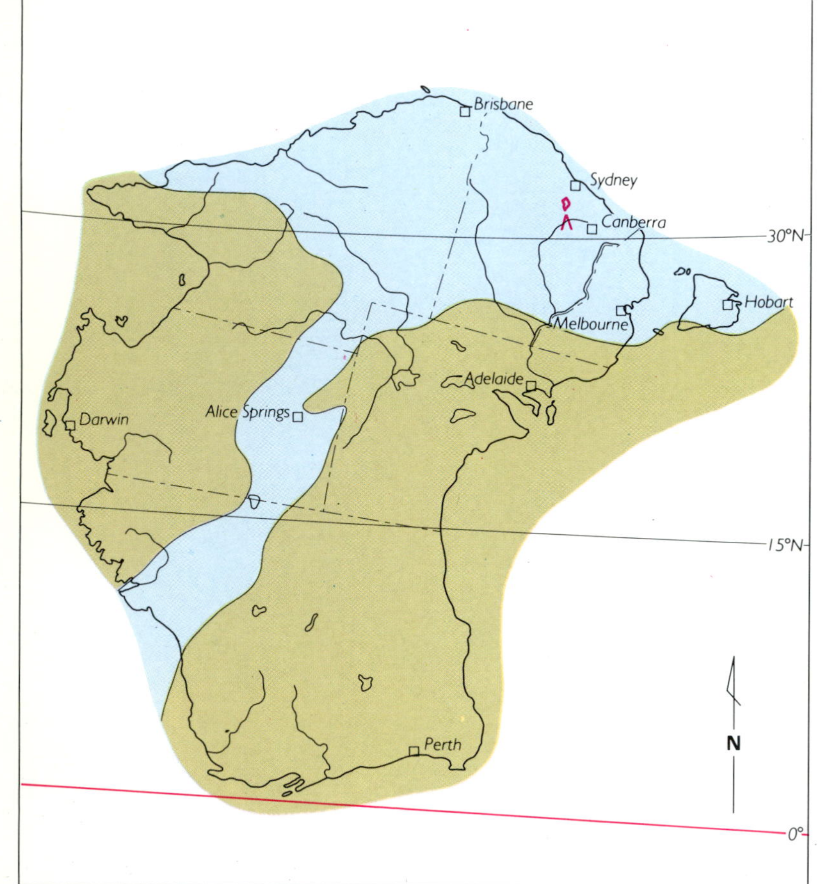

PALAEOGEOGRAPHY OF THE LATE ORDOVICIAN.

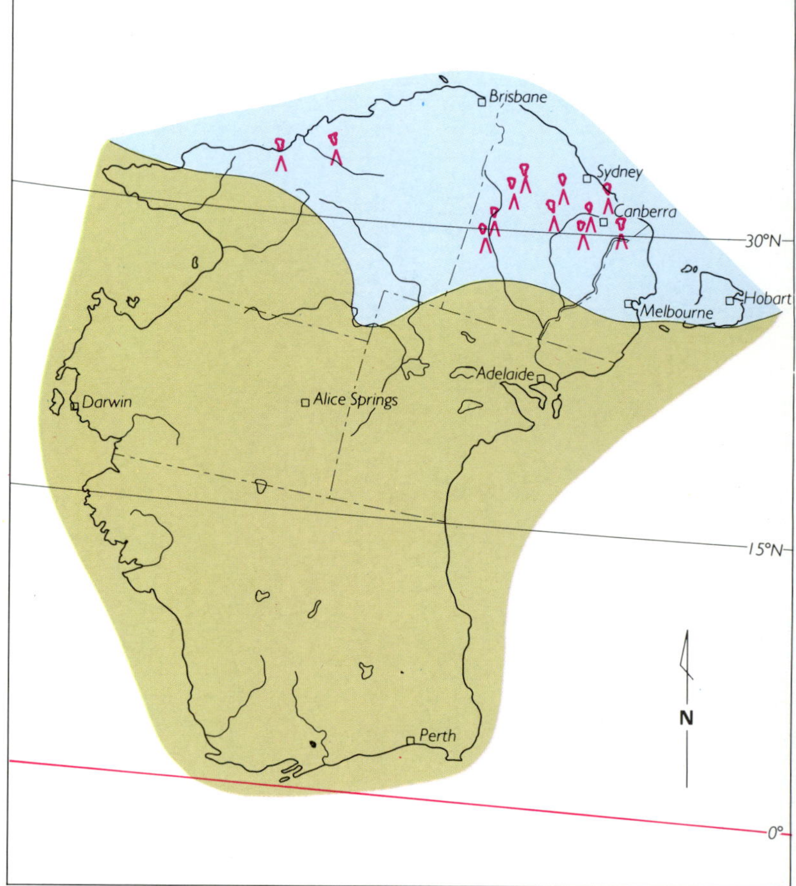

THE PHANEROZOIC EON

The 600 million years of the Phanerozoic Eon — from the Cambrian Period to the Present — represent only a seventh of the Time Column since life began. During this time, life as we know it has left a clear record of its existence in the form of fossils in sedimentary rocks.

The seas of Cambrian and Ordovician times teemed with animal life. The main Classes of Invertebrates were already established, and the first Vertebrates, the Fish, appeared in the Ordovician.

Animal life was evolving rapidly, leaving a comprehensive Fossil Record in marine sediments. Plants figure little in this record, although they must have been present in great quantity to feed the animal population. However, the soft tissues of Algae make them unsuitable for preservation as fossils, and only a few are found, whereas the shells, exoskeletons and bones of the animals were readily preserved.

By Late Cambrian times, 500 million years ago, the land which was to become Australia lay in the Northern Hemisphere. Part of the present-day coastline of Western Australia was on the Equator, and the rest of the continent lay between 0° and 30°N. At this stage the land area was only about half of the present-day continent, and an arm of the sea penetrated deep into Central Australia from the east. Later the sea retreated due to mountain building and general uplift in the Northern Territory. Formation of a zone of mountain building from western Tasmania, through western Victoria to north-central Queensland followed. 18

By Mid Ordovician times, 475 million years ago, the sea had entered the Canning Basin of Western Australia and flooded across the middle of the continent. Erosion was rapid during this period, as evidenced by the great volumes of sediment deposited. It is likely that the whole of the Earth was subjected to a high rainfall, and a hot climate, at this time.

During the Late Ordovician Period, about 440 million years ago, the seas regressed from the continent and there was uplift and erosion of the sedimentary strata which had been laid down in them in the Centre.

The remainder of the Late Ordovician and the Early Silurian continued to be times of geological unrest. The whole of Central Australia was being uplifted and rapidly eroded. There were mountain-building movements along the eastern continental margin.

In eastern Australia, offshore from a generally narrow shelf, elongated islands were formed, their positions commonly changing with time. It was in the almost enclosed sea between the margin of the continent and these islands on the shelf that sediments were collecting which were to form the Silurian rocks whose uppermost strata were to contain the famous *Baragwanathia* Flora of Victoria.

By the Late Silurian, 415 million years ago, life was ready to move onto the land. At that time the landmasses of the world are believed to have been more or less aggregated into the single supercontinent referred to as Pangaea. The Equator bisected Pangaea, passing across the top of the Australian region.

LEGACY OF A WATER ANCESTRY

Algae had by now reached a stage of considerable diversity. The time had come for life to move out onto the land. The three groups of Seaweed — the Green, Brown and Red Algae — were established and some had complex reproductive mechanisms. A number of these Algae have remained little changed from that distant age right up to the present.

Our interest lies in the Green Algae, which are the ancestors of all Higher Plants. It is necessary to have some understanding of their life cycles in order to understand the trends of evolution in land-plants.

Simple, unicellular organisms like Cyanobacteria and Amoebae reproduce by

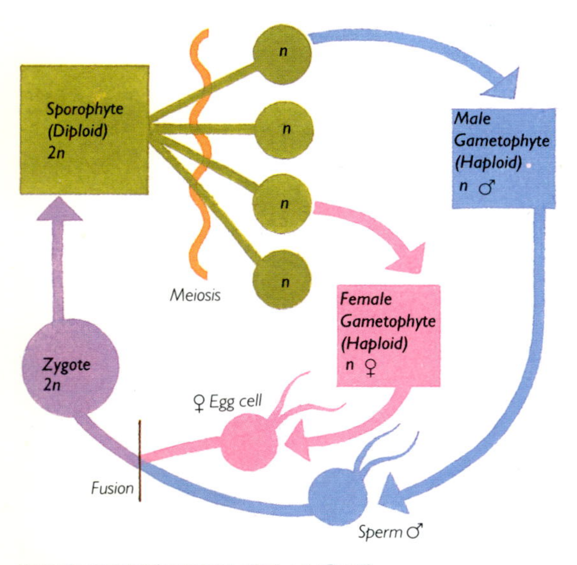

REPRODUCTION OF ALGAE

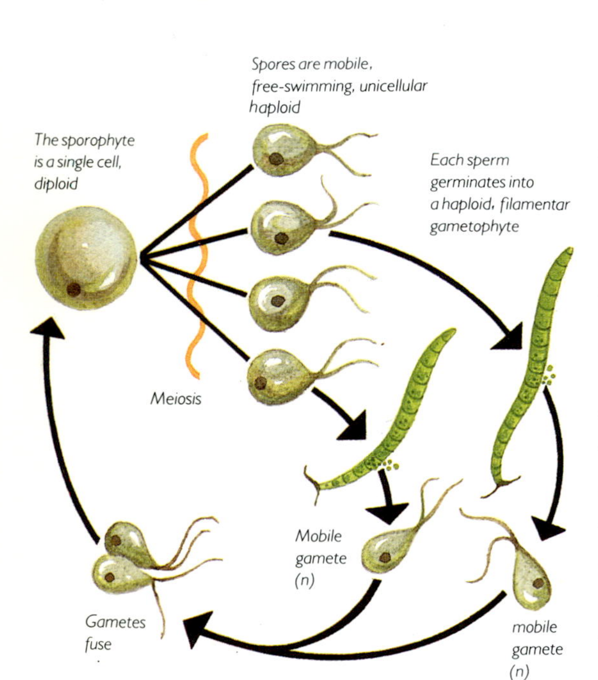

LIFE CYCLE OF *ULOTHRIX*

A simple Alga called Ulothrix illustrates the most primitive type of Algal life cycle. Here the sporophyte generation is unicellular, the gametophyte multicellular and filamentous. The gametes are identical and both are free swimming.

simple cell division. Asexual reproduction of this type results in each new individual being an exact replica of its parent. With the development of a more advanced cell type, which had a nucleus containing chromosomes that carried genes to determine the special make-up of each individual, sexual reproduction began to occur. The reproductive cells of two individuals combined and the offspring received genetic material from both parents. Different combinations of characters were thus possible, and change and evolutionary advancement were accelerated.

In the life cycles of Algae there is a pattern of asexual and sexual reproduction called "alternation of generations". At the asexual phase, a sporophyte plant has double chromosome numbers and is called "diploid". The mature plant produces spores by a special type of nuclear division called meiosis, halving the chromosome number. The spores are "haploid" and when they germinate they produce gametophyte plants, which produce male and female reproductive cells in special reproductive organs. Male sperm and female egg cells are known as "gametes". They fuse to form a "zygote" (with double chromosome number again), from which the diploid sporophyte plant develops.

The basic alternation of generations shows many variations in unicellular and multicellular Algae. The plants of the two generations may look the same or they may look different. The gametes may appear identical (both, say, may be mobile and the same size) or they may vary, with the female being larger and less mobile or even completely immobilised and retained within the tissues of the gametophyte to be fertilised there by the mobile male gamete. Because the male always has to swim to the female to effect fertilisation, water is the basic requirement for reproduction, which presented no problem while plants occupied water habitats. But when plants took to the land, difficulties arose.

The evolution of increasingly sophisticated reproductive mechanisms to eliminate the problems inherited from a water-dwelling ancestor is a major part of the evolutionary story of Higher Plants.

The sequence of evolution is from algal-type reproduction, through to more complex arrangements in other spore-producing plants, and then to seed formation without an alternation of generations. The gametophyte generation is progressively reduced to a few cells retained inside the sporophyte to produce the gametes.

The seed plants are of two kinds: those in which the ovule is "naked", the Gymnospermae; and those in which the ovule is enclosed in a vessel, the Angiospermae. (The names are derived from the Greek words *gymnos* meaning "naked", *angios* meaning "vessel", and *sperma* meaning "seed".)

The sequence from free-swimming gametes in a water habitat to Angiosperms with enclosed ovules held within an ovary until ready to germinate is paralleled in Vertebrate animal evolution: the Fish shed their eggs and sperm into the water to be fertilised and developed separately, and the life cycles are aquatic; then came the Amphibians, still dependent on water for the larval stages of their life cycle (e.g. eggs and tadpoles confined to the water, and Frogs and Toads terrestrial); then followed the Reptiles, which perfected the egg and cut the ties binding them to the water; and finally came the placental Mammals, in which the young are retained inside the mother's body until ready to start life on their own.

One can equate the Gymnosperms with the Reptiles and the Angiosperms with the Mammals. That the Flowering Plants and the Mammals came to dominate the world floras and faunas at the same time is cause for some philosophical satisfaction. It had taken land-plants 300 million years to fulfil the inbuilt genetic drive towards this point and, bearing in mind the common ancestry of plant and animal life in the first organisms right on the borders of life itself, it is no surprise that they reached the same point in evolution at the same time.

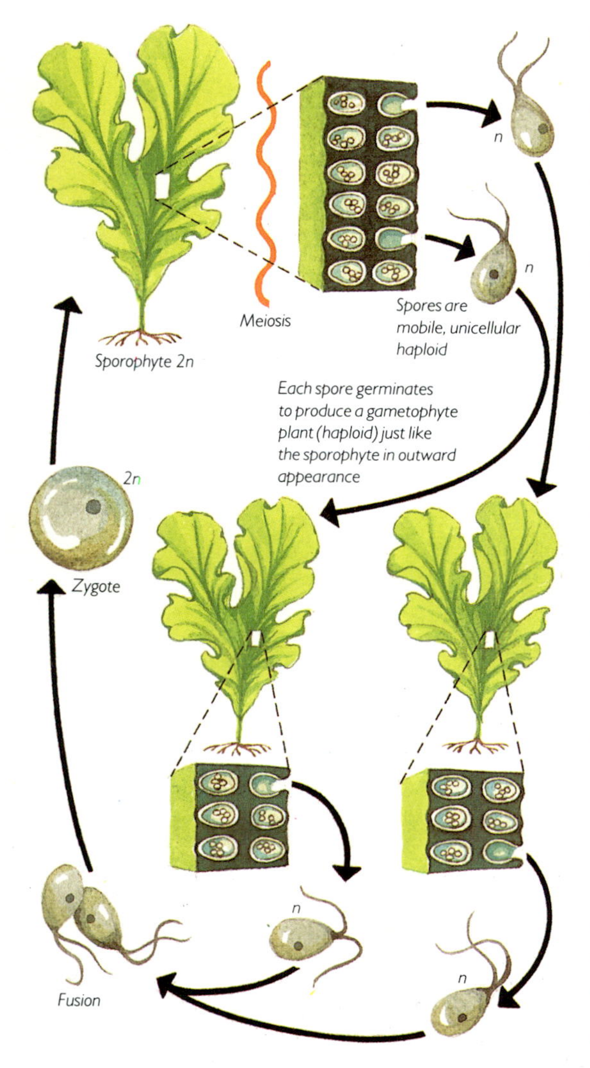

LIFE CYCLE OF *ULVA,* THE SEA LETTUCE

In the more advanced and complicated Sea Lettuce called Ulva the two generations look the same, but the cycle of haploid and diploid generation still applies.

Just as reproductive mechanisms had to evolve to suit life on the land, the plant body had to develop adaptations to suit a land habit.

The algal plant body as seen in Seaweeds is composed of a dense mat of filaments and there is no differentiation of tissues comparable with that in Higher Plants. The weight of the plant is supported by water and there is no need for strengthening of tissues to support it. Absorption of water can occur through all cell walls all over the plant surface. Drying out is not a problem, except for a few hours in the intertidal zone, but even here it has been simply solved by secretion of mucilage.

It is immediately apparent that life on the land imposed a whole set of new conditions which had to be met by evolutionary changes if land-plants were to succeed. How they met the challenges of a new environment is the story of evolution of a land flora, and it is told by the plant fossils of the different geological Periods.

The first plants on land were unicellular or thread-like Algae forming a green scum in swampy places. An association of such Algae with primitive Fungi and Bacteria, to break down and recycle dead organic matter and thereby help to mobilise soil nutrients, resulted in the creation of soil from the barren sand. A fringe zone suitable for permanent occupation by a land flora was thus created.

The symbiotic relationship between Fungi and Algae, and that between Fungi and primitive land-plants at the beginning of life on the land, would have been of the very greatest significance. In the water environment, the nutrients are dissolved and simply have to be absorbed. Transition to the land without the aid of Fungi would have been extremely difficult, for the first land-plants lacked specialised structures (i.e. roots) to absorb water and nutrients.

Modern studies on Soil-fungi show that these simple, almost structureless direct descendants of the earliest primitive Fungi play a very special role in the colonisation of barren sand. Mycorrhizal Fungi spread out through the surface layers in [19] barren sand dunes where the phosphorous and nitrogen compounds needed by green plants are held bonded to the silica grains and unavailable to plant roots. Being microscopic and thread-like, they wrap themselves around the grains and by their living activities free the nutrients and make them soluble and available to the plants colonising the dunes.

The first land "flora" can be visualised as symbiotic communities of Algae, Fungi and Bacteria learning to live together. As the Algae evolved into land-plants, the diffuse symbiotic relationships themselves evolved. The result was mycorrhizal infection of those parts of the plants being used for absorption. Bacterial arrangements for fixing of nitrogen or mobilisation of other elements for use by the plants were associated. That this is surely what must have occurred is strongly suggested by evidence of the incredibly high incidence of symbiosis between roots in living plants and Fungi. Old groups of plants like Cycads, Ferns and Conifers, which in the Australian flora represent "living fossils", all have symbiotic fungal relationships. More than 90 per cent of Angiosperms have mycorrhizal infections. Logic alone would imply that so fundamental a relationship as the symbiosis between Soil-fungi and plant roots must be a basic and very ancient arrangement.

There is good fossil evidence to support this theory. In the perfectly preserved, petrified Rhynie Flora of Scotland microscope sections of stems reveal fungal [20] threads and "vesicles" inside the cells. The Rhynie Flora is one of the oldest land floras, about 395-415 million years old, and a rare example of petrifaction of extremely delicate plants. When it was first described, scientists assumed that the Fungi had invaded the plant tissues after death and before fossilisation. It is now accepted that they are Mycorrhizal Fungi of the commonest "V.A." (vesicular-arbuscular) type which has worldwide distribution today.

CHAPTER 2

LIFE MOVES OUT OF THE WATER AND ONTO THE LAND

THE LATE SILURIAN TO EARLY DEVONIAN PERIODS

APPROXIMATELY 420 TO 385 MILLION YEARS AGO

Next to the evolution of life itself, the single most significant step in the evolutionary process was the adaptation of plants to life on the land. It was the step which enabled animals to leave the water and started the processes which eventually led to the origin of Man.

The first land to be colonised represents the first climatically transitional zone. The "climate" of the seas had been stable, constant and predictable. The climate of the land was the exact opposite. In this zone there were great pressures towards evolutionary change, resulting in the rapid development of new forms and their radiation out into fringe areas close to water.

Late Silurian and Early Devonian floras are characterised by plants which were small in stature, all probably less than a metre in height. They were not capable of "secondary growth" in their vascular systems and developed only as much conducting tissue as was laid down in the growing zones of the embryo. (Secondary wood production in trees and shrubs enables continuing patterns of growth and expansion.)

It is believed that the first land-plants were homosporous (i.e. their male and female spores were the same). This arrangement is the simplest in spore-producing plants.

Lycopodium phlegmaria, a modern Lycopod in which the fertile leaves (sporophylls) are different from the sterile leaves. Pendulous, forking branches terminate in branching strobili (cones) in which the fertile leaves are smaller than those which clothe sterile branches.

62 Baragwanathia longifolia, the tip of a young frond, from the Late Silurian of Victoria, age about 415 million years. (× 6.0)

In the Late Silurian, sediments accumulating in the shallow sea between the offshore islands and the eastern continental margin incorporated the earliest land floras. At this time, when life was taking the all-important step of leaving the water habitat, Australia was experiencing great tectonic upheaval. In south-eastern Australia, crustal movements of varying intensity affected the palaeogeography. It was only in the deeper troughs near Melbourne and west of Sydney that deposition of sediments was unaffected.

This geological unrest accounts for the paucity of the Fossil Record for this important stage in plant evolution. Intensity of crustal movements in one area resulted in newly formed Silurian rocks being folded, intruded by granite, and eroded before being covered by Early Devonian strata — which shows the magnitude of geological unrest at the time. In other places the mountain-building was accompanied by widespread volcanic activity and igneous rocks were being erupted into newly formed land areas.

63

63 Lycopodium squarrosum, a modern Lycopod very similar in appearance to Baragwanathia longifolia (the Lycopod which gives its name to the Baragwanathia Flora of Victoria and is one of the earliest land-plants, being dated at 415 million years in its oldest horizon). Lycopodium squarrosum is regarded as a primitive species as its fertile leaves (sporophylls) do not differ from the sterile leaves, the same condition as is seen in Baragwanathia.

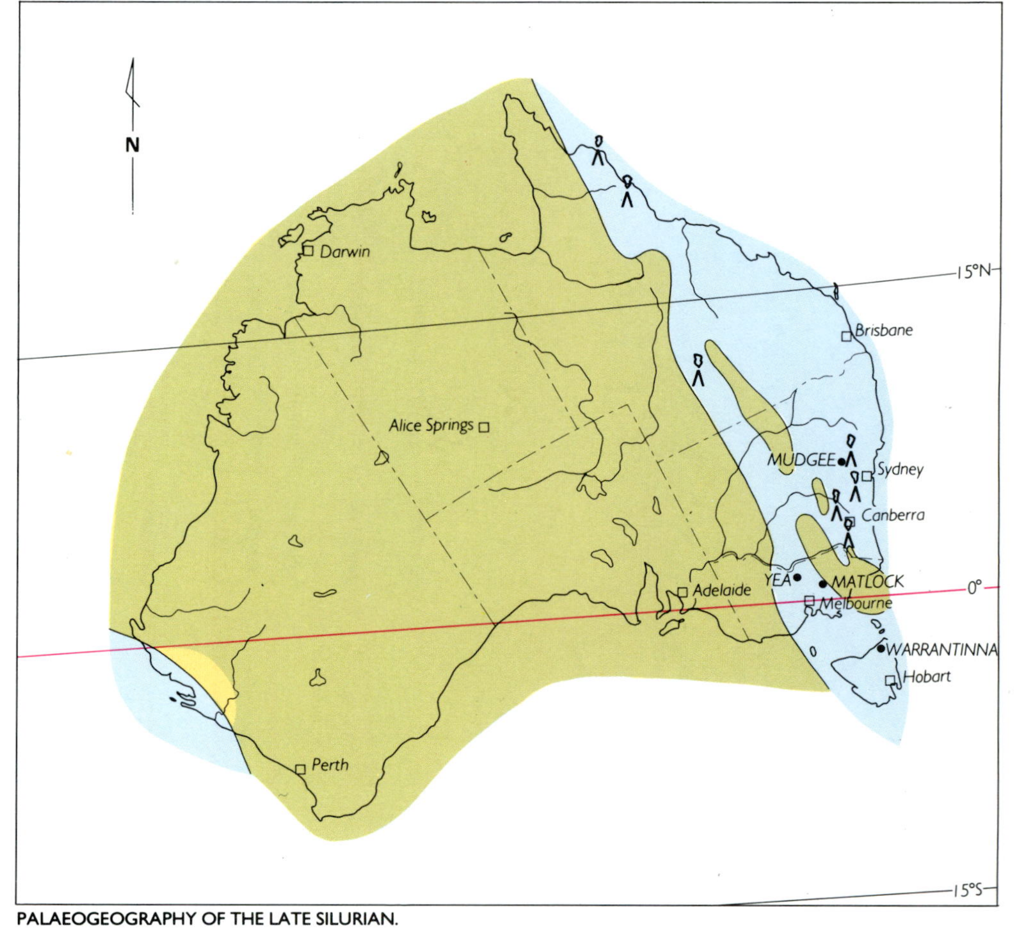

PALAEOGEOGRAPHY OF THE LATE SILURIAN.

PALAEOGEOGRAPHY OF THE EARLY TO MIDDLE DEVONIAN.

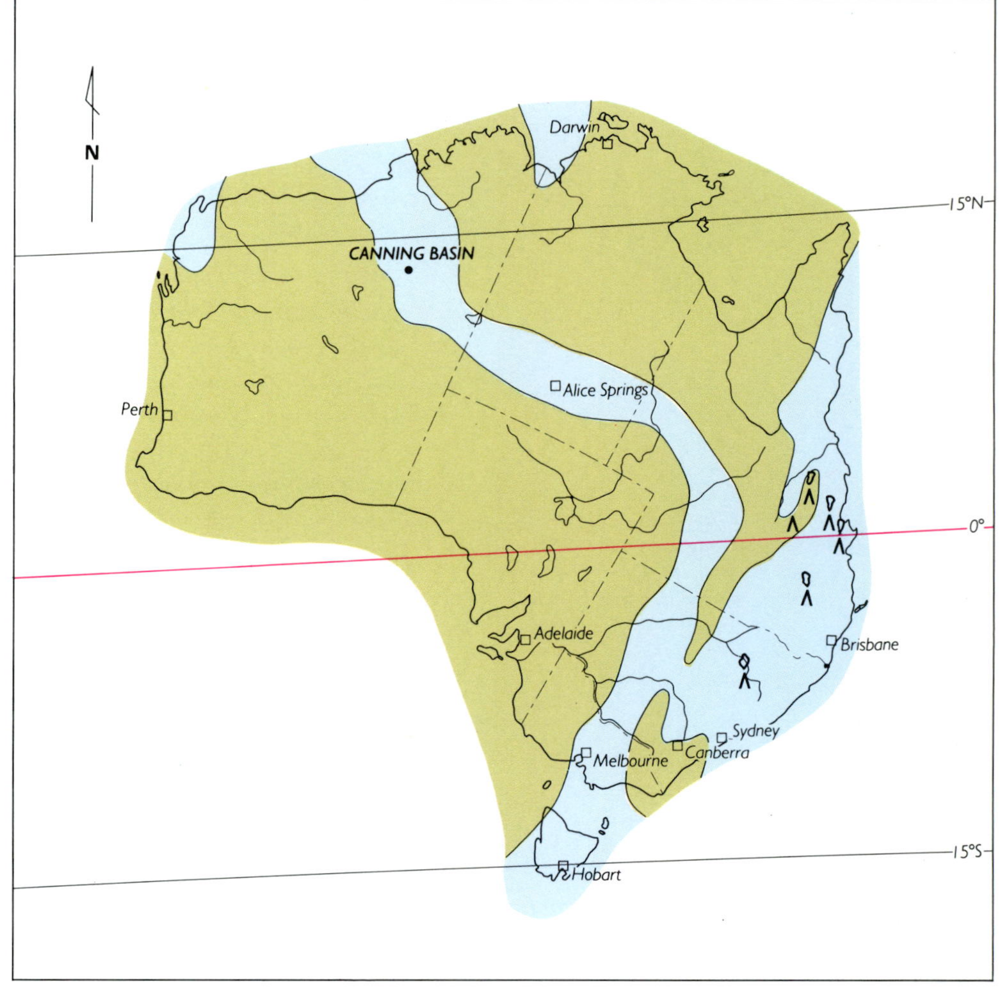

In Early Devonian times, the sea again invaded the Canning Basin of Western Australia and flooded across the middle of the continent. At first the sea was continuous from the western coast to the eastern seaboard. Then it started to retreat, leaving a line of small seas and finally disappearing. Sands of this sea, some windblown but most carried by rivers, accumulated across the area.

THE FIRST LAND-PLANTS

It is fortunate that the Green Algae represented a rich genetic pool when the time came to leave the water and start the greening of the Earth. The Algae's inbuilt genetic potential for evolution towards an increasingly immobilised female gamete (and the retention of the egg within the tissues of the plant) opened the pathways towards the production of seeds rather than spores and ensured the later arrival on the scene of the Gymnosperms and ultimately the Angiosperms.

The very nature of fossils makes it unlikely that the microscopic or even macroscopic but still delicate and almost structureless algal or algal-fungal plants of the first waves of land dwellers will ever be found to investigate. Some Lichen-like plant fragments found in very early rocks elsewhere in the world lend support to the concept of symbiotic relationships from the beginning.

Lichens are a symbiosis of fungal threads which form a leafy organ and Algae which live inside the tissues and produce the nutrients. They show great ability to colonise a wide range of the harshest habitats, from Arctic tundra to tropical desert where they may grow on granite boulders. While their constituent elements remain primitive and unspecialised, their adaptations to all types of habitat and their ability to deal with all the conditions encountered imply the conservatism of a very old life form.

64 H-branching of a Psilophyte, showing how the plant consisted of a mat of fine branches. From Einasleigh, Queensland, age about 395 million years. (× 2.5)

64

Structural alterations to the algal plant body to enable it to succeed in the land habitat were rapidly achieved. As the plant's weight was no longer supported by water it required strengthening tissues. With absorption of water and nutrients no longer a function of all surface cells special organs had to evolve, and thus roots were developed for this purpose and to anchor the plant. Conducting tissue had to be perfected to carry both water and the foods prepared in photosynthesis throughout the plant body. Production of specialised photosynthetic organs proved advantageous, and leaves were developed. As water loss from tissues exposed to the atmosphere was now a problem, a waterproof skin was developed for protection, and the waxy cuticle came into being. As there was need for gaseous exchanges during photosynthesis and other life processes, pores were required in this impervious cuticle layer, and thus developed the stomata, with specialised guard cells designed to control the size of the pores under different conditions.

VASCULAR PLANTS

Those plants which developed conducting tissue of specialised woody vessels (or xylem) to conduct water are known as "vascular plants". The xylem at the same time gave rigidity to tissues and enabled stems to stand erect and display their photosynthesising organs to maximum advantage. All land-plants today, except for Mosses and Liverworts, are vascular plants. 21

The oldest recorded Higher Plant that has left a fossil record was more like a Moss than a vascular plant. Called *Cooksonia* it occurs worldwide in Late Silurian strata. It has very thin, forking branches bearing terminal sporangia often no bigger than a pin-head, with spores with resistant covering enabling them to be dispersed by wind. 22 23

From a *Cooksonia*-like ancestor the first easily recognised vascular plants evolved. From the beginning there were two distinct lines of evolution: one leading only to the Lycopods (or Clubmosses), and the other to all the other groups of Higher Plants. 24

THE *COOKSONIA* FLORA

Reconstruction of *Cooksonia*

Australia's meagre *Cooksonia* Flora occurs in Siluro-Devonian beds at Mudgee, New South Wales, and in beds of the Mount Daubeny Group further west, near Broken Hill. Aggregations of "carbonaceous detritus", fragmental and almost microscopic, the fossils are most unremarkable as specimens and yet so important in the story of land-plant evolution. It is likely that many more localities will be discovered and more information will become available about these earliest land-plants. To date they have not been actively sought in Australia and, being easily overlooked in the field, geologists have not collected many samples.

The worldwide distribution of *Cooksonia*-type plants in Late Silurian sediments helps to confirm the existence of Pangaea. 25

A similar distribution is seen in the earliest Rhyniophytes and Zosterophylls which follow the *Cooksonia* plants. They occur from Australia to the Canadian Arctic, in the United States, Spitsbergen, Britain, Western Europe, Greenland and western Siberia. Sedimentological evidence showing correlation of stratigraphical layers and other features, which indicates common origin in one landmass, also confirms the concept of Pangaea during this period. 26 27

CONTINENTAL REASSEMBLY FOR MIDDLE SILURIAN.
424 million years ago.

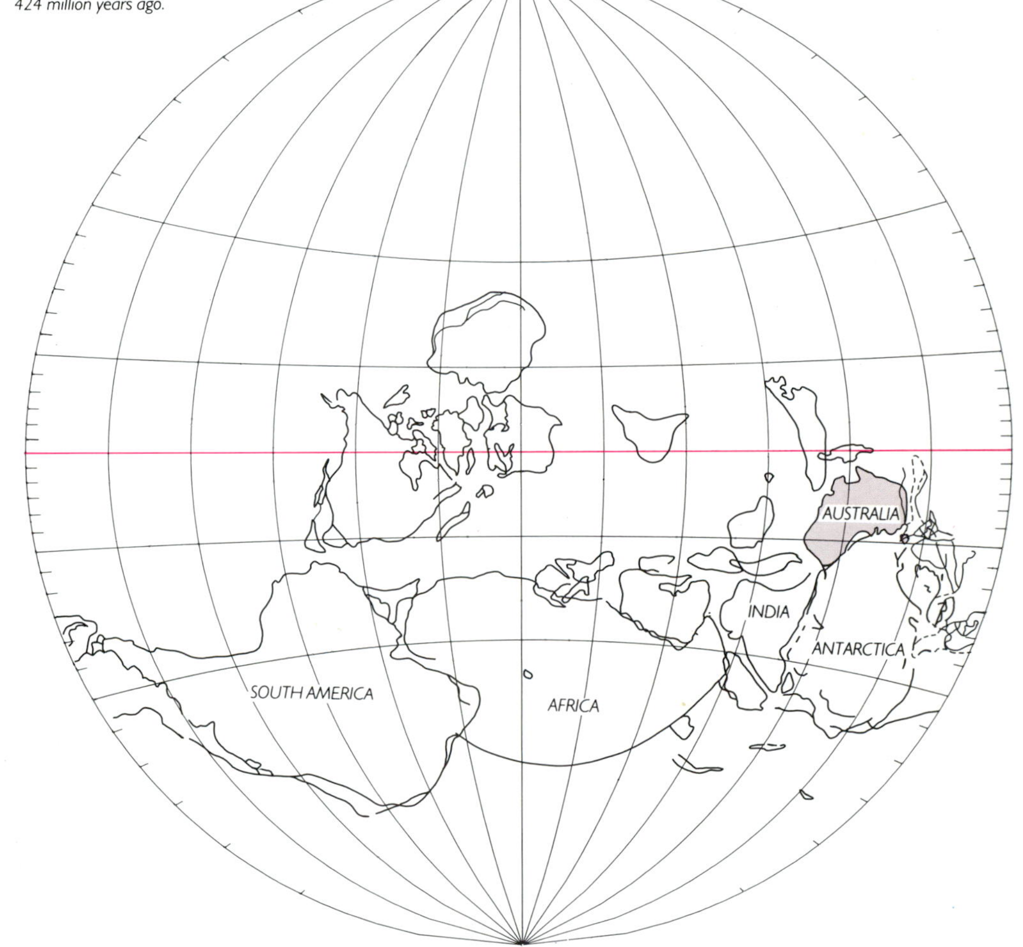

65 *Cooksonia fragments. Carbonaceous detritus with fine stems and pinhead-sized sporangia (photograph much enlarged). From Mudgee, New South Wales, age about 415 million years. (× 2.5)*

65

ZOSTEROPHYLLS, THE ANCESTRAL TYPE FOR THE CLUBMOSSES

Like *Rhynia*, Zosterophylls are known in complete detail because of their presence in the Rhynie Flora. They are ancestral to the Clubmosses, which were dominant in early Palaeozoic floras.

In this group, leaves are small scale-like appendages covering the stems. They started out as small projections which effectively increased the photosynthetic area, and as they became enlarged and flattened they received a branch from the conducting tissue of the stem. The sporangia of Zosterophylls are borne laterally, in the angles of leaves, not terminally as in *Rhynia*.

The importance of this group of plants declined after the Palaeozoic Era and today there are only four genera of living descendants: *Lycopodium*, *Selaginella*, *Phylloglossum* and *Isoetes* (the Quillworts).

Reconstruction of *Zosterophyllum*

FAMILY TREE OF EARLY LAND PLANTS

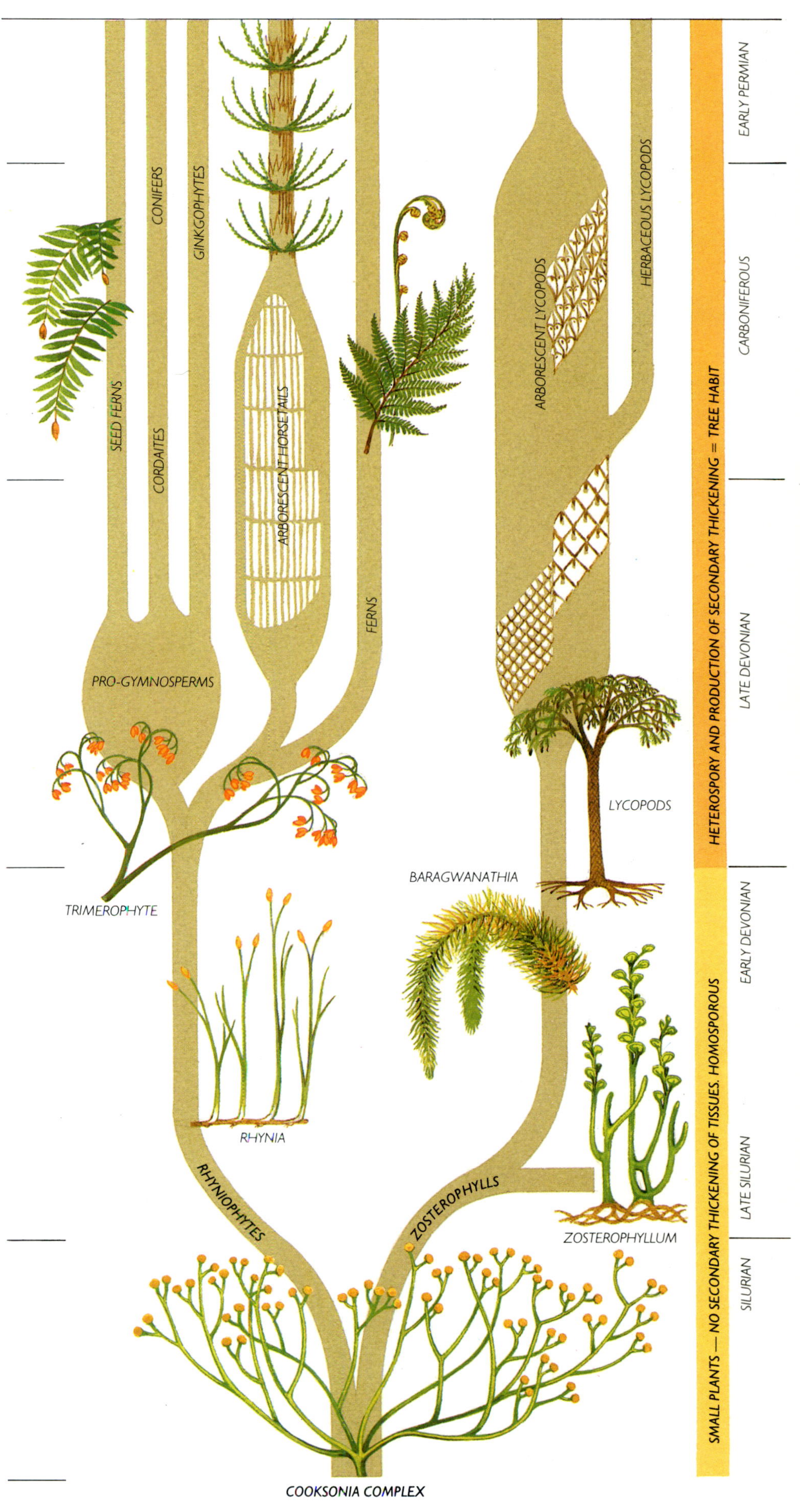

Reconstruction of *Rhynia*

RHYNIA, THE ANCESTRAL TYPE FOR FERNS, HORSETAILS AND SEED-PLANTS

We know a great deal about this early vascular plant because it is found in the Rhynie Flora of Scotland — a famous flora occurring in a deposit of silica rock (chert) at Rhynie in Aberdeenshire. Here, the delicate early land-plants grew in a marshy site. Volcanic eruptions nearby flooded the marsh with boiling, silica-rich water. The plants were killed and the silica permeated their tissues, preventing decay and replacing plant molecules with mineral matter, resulting in preservation of the finest microscopic structure.

Rhynia plants were leafless, composed of small erect branches arising from networks of prostrate branches which absorbed water and nutrients from the substrate, probably aided by Mycorrhizal Fungi. They had stoma on their cuticularised surfaces and they bore terminal sporangia. Photosynthesis was carried out by all green cells, anywhere on the plant. The stems had a core of vascular tissues with woody xylem "tracheids" (elongated elements designed for water transport) and special conducting cells called the "phloem" to transport prepared foods.

Both the sporophyte and the gametophyte generations of the plant are known as petrifactions — the sporophyte bearing the terminal sporangia, and the gametophyte generation of similar stems which bear sperm and egg cells in little cups in their surface tissues. It may be that the gametophyte had only prostrate stems while the sporophyte had erect stems for raising the spores above the ground to aid distribution.

The evolution of leaves on the *Rhynia* line is believed to have been by modification of small lateral branches, which were the product of unequal divisions of the stem. These branches were later flattened to produce a larger photosynthetic area. The addition of a webbing of tissue between the small, flattened branches resulted in the production of a leaf lamina.

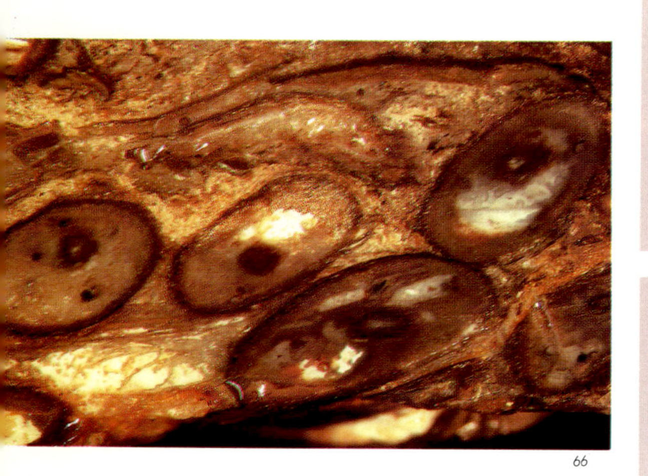

66

66 A slice of Rhynie chert, showing stems in cross-section. The largest stems are 2 mm in diameter. The Rhynie fossils are rare examples of very delicate plants preserved as petrifactions with cell structure visible under a microscope. They are among the earliest land-plants in the Fossil Record, and a knowledge of their anatomy has been vital to the understanding of plant evolution. Specimen from Rhynie, Aberdeenshire, Scotland. Age about 410 million years.

MODERN RELATIVES OF *RHYNIA*

Compare the living *Psilotum* and *Tmesipteris* with the first land-plants . . . so little change in 400 million years!

The *Psilotum* sporophyte plant is rootless, with forking rhizomes and stems. It has spirally arranged lateral appendages that are scale-like or leaf-like. A core of xylem acts as the conducting strand in the axes. The sporangia are thick-walled and homosporous, and are borne at the end of very short, lateral branches.

Psilotum's colourless rhizomes have rhizoidal hairs, and function as organs for absorption. Mycorrhizal Fungi gain access to the cortical cells of the rhizomes through the rhizoids. (In the absence of true roots, this primitive association enables the plants to exist and function in quite arid environments. *Psilotum nudum*, for example, grows in crevices in the Hawkesbury Sandstone outcrops in the Sydney region of New South Wales.)

Some branches of the rhizome turn upwards and develop into aerial shoots which are green and bear minute appendages. These "leaves" have no vascular tissue. The stems have stomata and are the main photosynthetic organs. The axes branch in a regular dichotomous manner (equal forking) and the distal regions are triangular in cross-section. In the upper regions of the more vigorous shoots, the "leaves" are replaced by fertile appendages which are short lateral branches each bearing two "leaves" and terminating in three fused sporangia.

The gametophyte is like a piece of sporophyte rhizome (as was the case in *Rhynia*) consisting of branching axes without chlorophyll. Sex organs borne together on the same gametophyte produce the mobile gametes.

Tmesipteris is more highly evolved in respect of its leaves. Those on the rhizome and stem base are scales. The main frond has leaves with stomata, and there are no stomata on the stem. These plants are epiphytic on Tree-ferns, and the gametophyte is the same as in *Psilotum*.

67

68

THE *BARAGWANATHIA* FLORA OF VICTORIA

This flora, comprising *Baragwanathia* and associated plants, is one of the most important ancient vascular land-plant floras of the world. It is one of the best preserved, and one of the richest in terms of the number of different plants which it contains. It has excited great scientific interest since its discovery at Yea in Victoria in 1875 and its subsequent description in 1935 by Australia's eminent pioneer palaeobotanist, Dr Isabel Cookson, working in conjunction with Dr Lang of the United Kingdom. 28–35

Baragwanathia is now recognised as being a Lycopod, derived from the Zosterophylls. It is similar in organisation and structure to living Lycopods, especially *Lycopodium squarrosum*. Because of its high degree of specialisation, there has been heated controversy over the age of the beds in which it first occurs. The sequence of strata is continuous from the Silurian to the Early Devonian. At Yea *Baragwanathia* occurs with Rhyniophytes and a Zosterophyll, and with Graptolites (which are Invertebrates). The Graptolite has been used to correlate the fossil horizon with the Ludlow Division of the Silurian in Wales and elsewhere in the world. Until recently, only poorly preserved examples of the Graptolite were found and the correlation was considered to be dubious. Recent evidence from the study of excellently preserved examples confirms the Late Silurian age for the Yea locality. 34 35

The very advanced appearance of *Baragwanathia* compared with that of the Zosterophylls from which it has evolved was the cause of the scepticism about the Late Silurian date. It is, however, becoming clear with the increasing volume of information on land-plant spores and fragments in pre-Late Silurian rocks that the

69 *Baragwanathia longifolia*. Apex of a mature frond. Age 395 million years.

67 *Psilotum nudum*, a living *Rhynia*-like plant. Its green, forking stems are the photosynthetic organs. Very small spine-like projections (leaves) can be seen on some nodes. The sporangia are fused in groups of three at the ends of short lateral branches, each of which bears two tiny leaves. Its underground rhizomes have rhizoids, and no true roots, and Mycorrhizal Fungi help in the absorption of water and nutrients from the soil. The plant differs very little from its ancestors of 410 million years ago. It is widespread in the tropics and sub-tropics, and is cosmopolitan. *(Peter Valder)*

68 *Tmesipteris tannensis*, another relative of the Rhyniophytes, is confined to Australia, New Zealand and some Pacific islands. It grows as an epiphyte, mostly on Tree-fern trunks but occasionally on forest tree trunks and very rarely on the ground. This example is on the caudex of *Todea*, a Tree-fern growing on Mount Wilson, New South Wales. The leaves are much more developed than in *Psilotum* and are the photosynthetic organs. Fertile appendages (short branches terminating in two fused sporangia) replace some leaves near the tip of the stem. *(Peter Valder)*

POSITION OF AUSTRALIA IN RELATION TO OTHER LANDS IN THE LATE SILURIAN. *415 million years ago.*

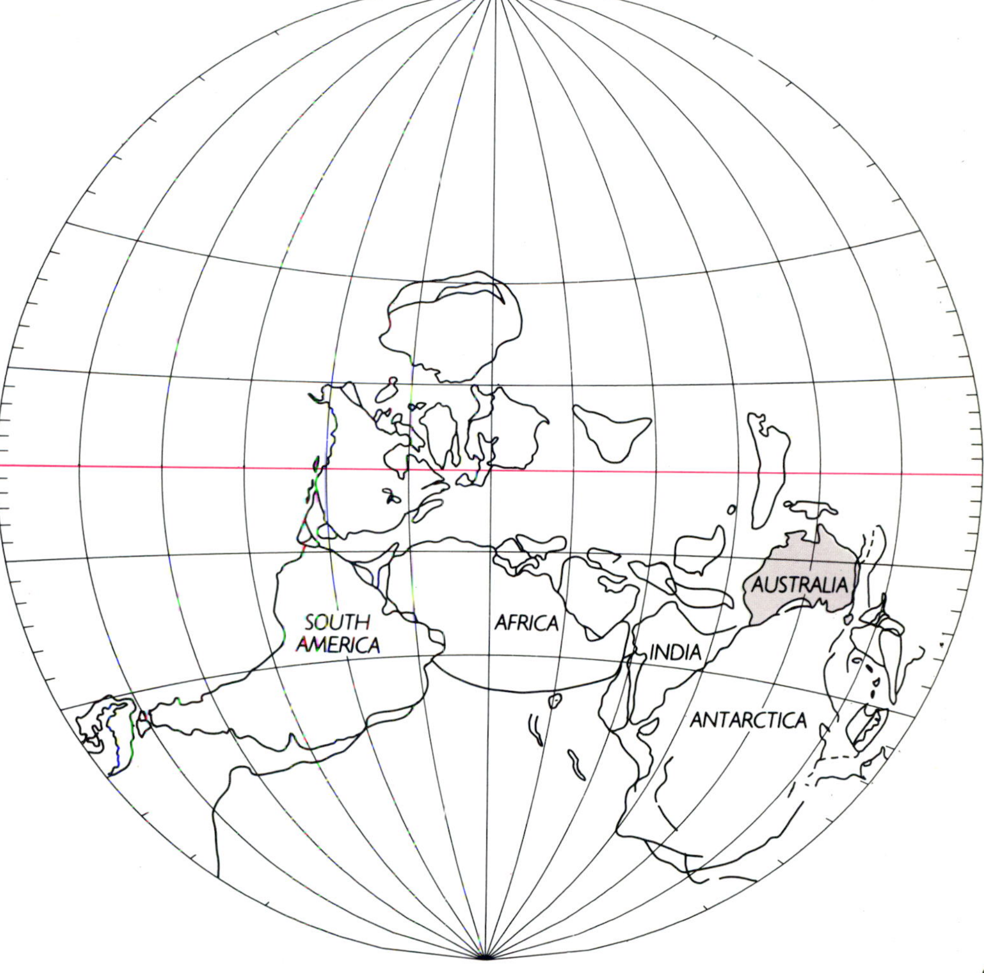

The similarity of Baragwanathia longifolia to modern Lycopods, particularly to Lycopodium squarrosum is striking. The Lycopods appear to have changed little over the vast period of time that they have been on Earth, when we compare the first with the most recent. But the Fossil Record shows that the early herbaceous forms were followed by Giant Clubmosses, far removed from our modern representatives.

vascular plants may indeed date back further than had been imagined. This greater age would render the degree of specialisation reached at such an early date less surprising.

The lower horizon for *Baragwanathia* at Yea, now accepted as Late Silurian, is separated by 1700 metres of siltstone from the upper assemblages which are of Early Devonian age. The deposition of this depth of siltstone occurred during a period of about 30 million years and there are many more plant species in the upper flora. Trimerophytes, for example, which evolved from Rhyniophytes and belong to the group of early plants from which all the Higher Plants evolved, are present. 36

The *Baragwanathia* Flora is closely related to the late Silurian and Early Devonian floras of the rest of the world. As the concept of one landmass, Pangaea, has been established for these times, this similarity is not surprising.

70

70 Baragwanathia longifolia. Stems which have lost their leaves and show a patterning of the surface with spots indicating where leaves were attached. Age 395 million years. (× 0.7)

71 Baragwanathia longifolia. A forking stem clothed in long leaves. (× 2.0)

72 A young, fertile frond of Baragwanathia longifolia. The stem is densely clothed in leaves, and the sporangia are seen as circular lumps, particularly near the top of the loop in the frond. This specimen is from Yea, Victoria, and is from the lower horizon at which the species occurs, being of Late Silurian age, about 415 million years. (× 3.1)

71

73 Hedeia sp., a group of sporangia at the apex of an upright stem. From Wilson's Creek shales, Victoria. Age Early Devonian, about 395 million years. (× 3.7)

74 Pluricaulis biformis, a Zosterophyll fructification which has lateral sporangia, a characteristic of the group. Several sporangia, spirally arranged, are attached to the top half of the stem. From Yea, Victoria. Age about 415 million years. (× 3.8)

75 Yarravia sp., a slender erect stem with a group of terminal sporangia. From Yarra Track, Victoria. Age Early Devonian, about 395 million years. (× 3.3)

Pluricaulis, Hedeia and Yarravia erect fertile branches were borne on leafless plants which forked and branched forming a mat on the ground. The plants were vascular, with a core of vascular tissue in the stems and no true roots. There were rhizoids on the prostrate stems to absorb water and it is probable that all had a symbiotic relationship with Fungi to assist in water and mineral uptake. 36

73

74

75

76 *Dawsonites racemosa.* This simple Zosterophyll from Bowning, New South Wales, was associated with a species of Trilobite (an Invertebrate), which fixed its age as Early Devonian. The slender stem shows H-branching at the base, and the sporangia are lateral. Tracing shows the structure of this early plant more clearly. Age about 395 million years. (× 1.6)

77 *Palaeostigma sp.* A stem showing fine, regular pitting, and small, thin leaves attached to edges. It is an early Lycopod, age about 395 million years (Early Devonian) from Mudgee, New South Wales.

76

77

CHAPTER 3

THE GIANT CLUBMOSS FLORA

THE MID TO LATE DEVONIAN AND EARLY CARBONIFEROUS PERIODS

APPROXIMATELY 385 TO 325 MILLION YEARS AGO

By Mid Devonian times, land-plants had achieved two great evolutionary advances. The first related to their structure, the second to their reproduction.

The earlier land-plants were all small, their size limited by an inability to produce secondary wood. Now, the development of a cambium — a layer of cells which produces continuous layers of xylem cells to increase the girth of the conducting cylinder — provided a major breakthrough. It also produced new phloem, providing extra conducting tissue for transporting prepared foods, a necessary addition in larger plants.

Thus plants were no longer limited in size and could grow into trees. They could compete for sunlight and space by overtopping their rivals. Strong, woody plants had the added advantage of being more resistant to natural forces and to the predations of animals which depended on them for food.

Since increased size meant that reproductive organs were further from the ground, spore dispersal by wind was improved. But the free-swimming gametes' requirement for water in which to complete life cycles was a problem. Heterospory, retention of the female egg cell in the sporophyte, and the series towards seed formation, were responses to these problems.

After the Early to Mid Devonian times of major crustal movements, the Australian landmass entered a less hectic period of its geological history. General uplift continued, with the formation of high ground in Central Australia, and by the end of the Devonian Period mountainous terrain in the region now occupied by the Macdonnell Ranges had been formed. The shallow seas retreated generally eastwards. Major east-flowing river systems spread sediments over large areas, resulting in terrestrial sedimentary deposits.

Fish abounded in rivers and lakes, no doubt in response to the availability of food from thriving plant communities.

There was a global rise in sea level in the Mid Devonian, the cause of which is unknown. World climates at the time were warm and wet, and there were no polar ice caps. Eustatic sea level rises have usually resulted from melting of ice caps, an explanation not possible in this case. The rise in sea level caused only limited marine incursions in Western Australia and Victoria. Most of the continent was sufficiently uplifted to prevent extensive shallow sea formation.

By the Late Devonian, the Australian land area had moved through 35° of latitude from the position it had formerly occupied during the Upper Silurian to Lower Devonian Pangaea. It continued to move gradually southwards during the Early

A modern Horsetail, Equisetum sp., the only living genus of Sphenophyta, a Division of plants which was important in the Fossil Record from Devonian times. Giant Horsetails grew with Giant Clubmosses in the coal swamps of the Carboniferous of Europe. Their size decreased through time and now all the 25 species which exist today are perennial herbs. The green, segmented stems, the leaves in small leaf-sheaths at nodes and the terminal cones of modern species are of the same appearance as the herbaceous Horsetails so common in the Fossil Record. (Jim Frazier)

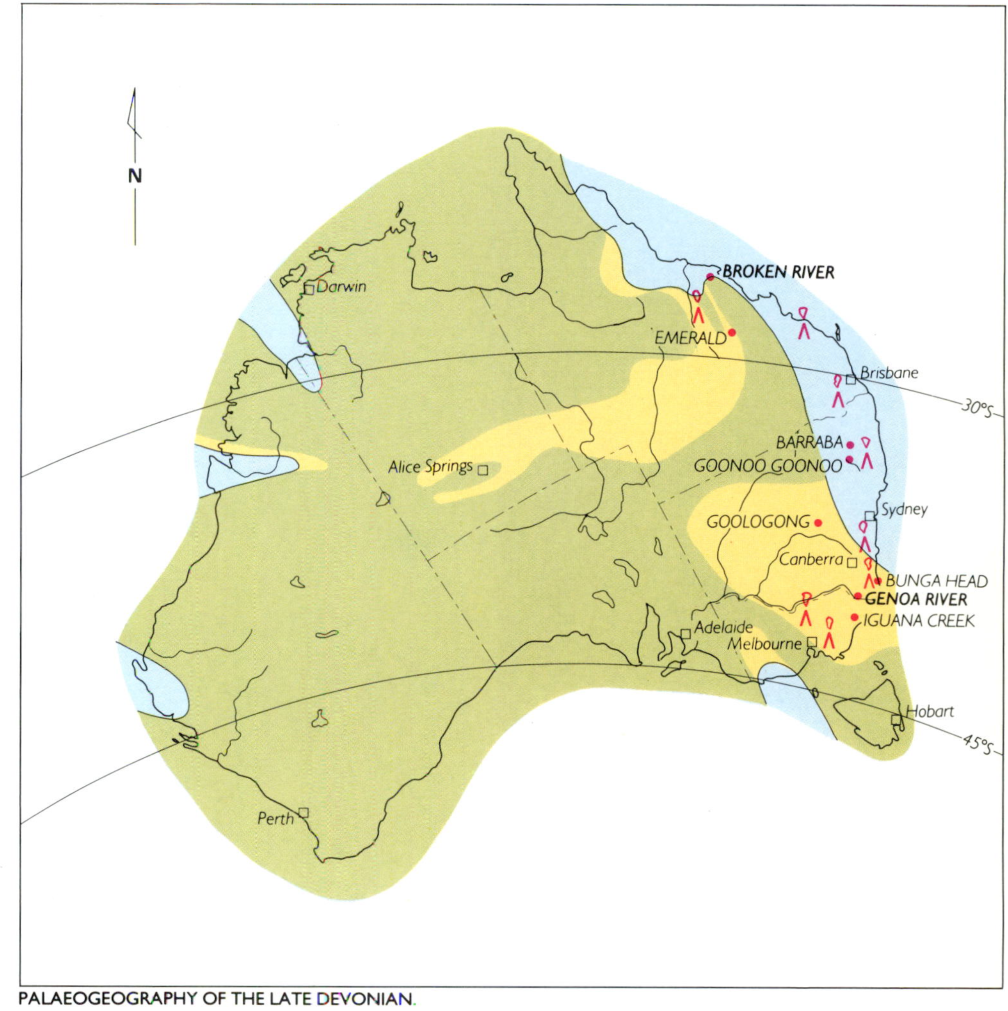

PALAEOGEOGRAPHY OF THE LATE DEVONIAN.

PALAEOGEOGRAPHY OF THE EARLY CARBONIFEROUS.

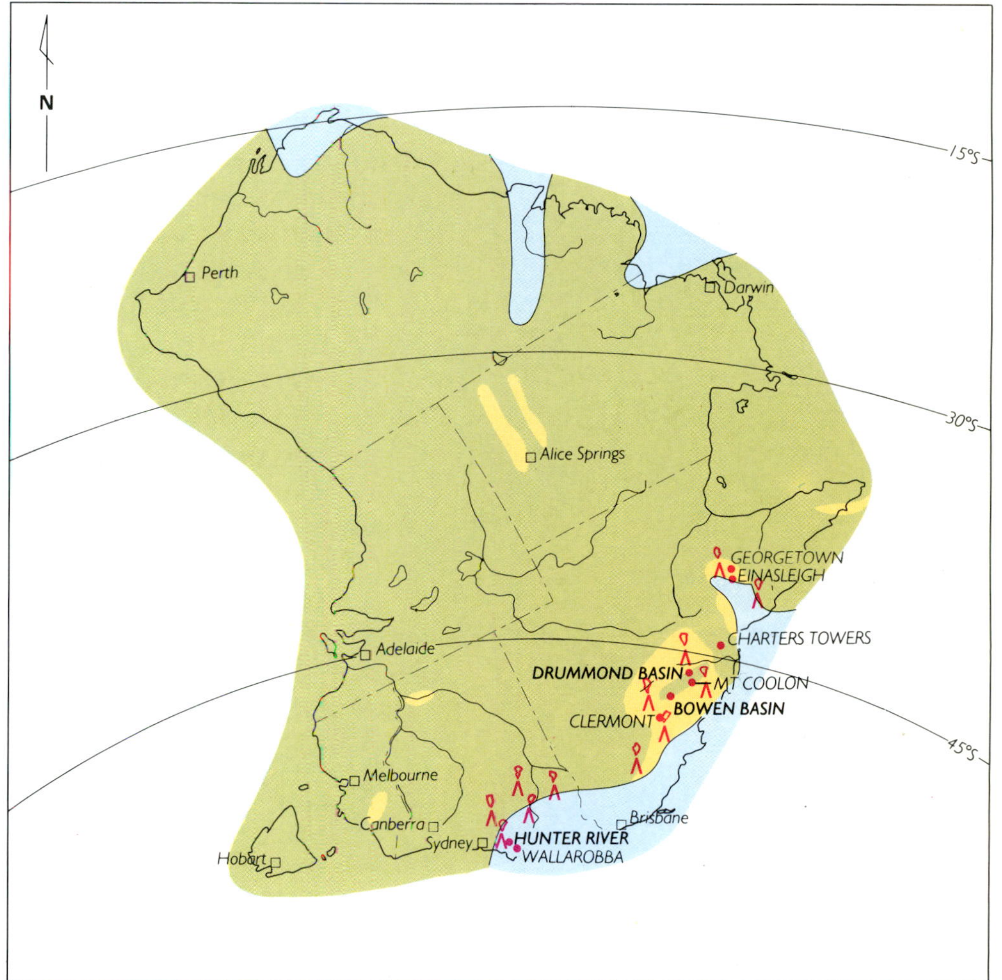

78

78 A *Lepidodendron* stem with branch scars, known as "*Ulodendron*". The specimen is from the Early Carboniferous at Wallarobba, New South Wales. The *Lepidodendron* flora, which attained such luxuriance in the tropical swamps of the Northern Hemisphere in Carboniferous times is not as luxuriant in the Southern Hemisphere, and in Australia the Giant Clubmosses are much smaller, though still giant when compared with modern herbaceous representatives of the Lycopods. The increasing cold as Australia moved towards the South Pole, and as the Late Carboniferous ice age approached, made most forms extinct, and the *Lepidodendron* flora is characteristic only of the Early Carboniferous. (× 1.1)

79 Part of an elongated *Lepidostrobus* cone from Mt Coolon in the Bowen Basin, Queensland. Some cones of this sort produced by the Giant Clubmosses in Europe are known to have reached 50 cm in length and to have contained billions of spores. It is unlikely that such long cones occurred in Australia, where the Clubmosses did not attain such great size and luxuriance, but this example is obviously of a far more elongated type than that illustrated in picture 80. Its sporophylls are very closely packed. (× 0.7)

80 This elongated cone has a small pedicel at its base, and there are bract-like sporophylls which have an acute bend clearly seen at its base. Called *Lepidostrobus*, the cones of *Lepidodendron* and related genera bore very large numbers of microspores if male, and a smaller number of megaspores in the female. This specimen is from the Early Carboniferous in the Clermont District of the Bowen Basin, Queensland. Age approximately 340 million years. (× 2.1)

CONTINENTAL REASSEMBLY FOR THE LATE DEVONIAN.
370 million years ago.

Carboniferous times when the Giant Clubmoss Flora was dominant. The end of that "*Lepidodendron* Flora" coincides with the "breakaway" when the continent was swept off to South Polar regions and rotated through 90°.

There was a great deal of volcanic activity on the continental shelf along the eastern margin of Australia during the Late Devonian and Early Carboniferous. Much of the sediment which accumulated on the shelf between the line of volcanoes and the Queensland coast was derived from these active volcanoes.

While the Early Devonian flora of primitive land-plants showed close relationships with world floras of the same age, the degree of correspondence decreases as the Australian flora becomes more highly evolved towards the Early Carboniferous. This decrease is a reflection of the opening up of the eastern side of Pangaea and the development of southern and northern regions (forerunners of Gondwana and Laurasia). Australia was increasingly isolated from the tropical, equatorial North American and northern European parts of Pangaea where the Giant Clubmoss Flora was best developed. In those regions it was responsible for the formation of the vast coal deposits of the Northern Hemisphere. There was no luxuriant expression of this flora in Australia, nor any coal formation during Carboniferous times.

Climate as well as geographical separation affected the development of the Giant Clubmoss Flora in Australia. The world was undergoing gradual climatic change towards a cooler phase which was to terminate in an ice age in the Late Carboniferous. Australia's drift southwards away from the Equator would have increased the effects of the general cooling.

Australia's Early Carboniferous plants are closely related to South American floras of the same age, particularly those of the Argentine. Relationships to the Northern Hemisphere plants are not as close. Again, this situation is an expression of climate: the Argentine and Australia were in similar latitudes.

By Mid Devonian times we find that the main groups of Higher Plants were already delineated. The spore-producing plants were the Lycopods (Clubmosses), Articulates (Horsetails) and Ferns. The seed-producing plants were the early Gymnosperms, the Seed-ferns and the pro-Gymnosperms.

LYCOPODS OF THE GIANT CLUBMOSS FLORA

37–40

The Giant Clubmosses have been reconstructed from suites of fossils which show their different parts. In *Lepidodendron*, for example, trees consisted of an unbranched trunk which forked repeatedly into a crown at a considerable distance above the ground. The smaller branches were clothed in leaves which were attached to leaf-cushions arranged in an ascending spiral, so that leafy twigs were like bottle-brushes. *Lepidodendron* leaves are assigned to the form-genus *Lepidophyllum*. After the leaves had fallen, the structure of the leaf-cushions continued to grow with the secondary growth of the branch, so that the pattern of leaf bases expanded. The outside of all branches, large and small, was thus patterned with ascending spirals of leaf-base scars, resulting in a scale-like sculpturing.

The pattern on the bark of individual trees varied, depending on the age and the size of the branches. The pattern varies further in fossils, because the impression preserved may represent the internal cast of a stem or a bit of shed bark, or the external cast of a stem in which outer layers had been removed prior to fossilisation. Patterns may also be distorted by compression. The leaf-bases show the position of the leaf-traces — the conducting strands to the leaf from the xylem of the stem.

Lepidodendron had cones comprising scale-leaves (sporophylls) which carried the sporangia. Male sporangia contained large numbers of small spores, while female sporangia had a few larger spores. The cones are called *Lepidostrobus*. It has been calculated that some of the trees in the coal swamps in Europe, which reached a height of 45 metres and had cones up to 50 centimetres long, must have produced

79

80

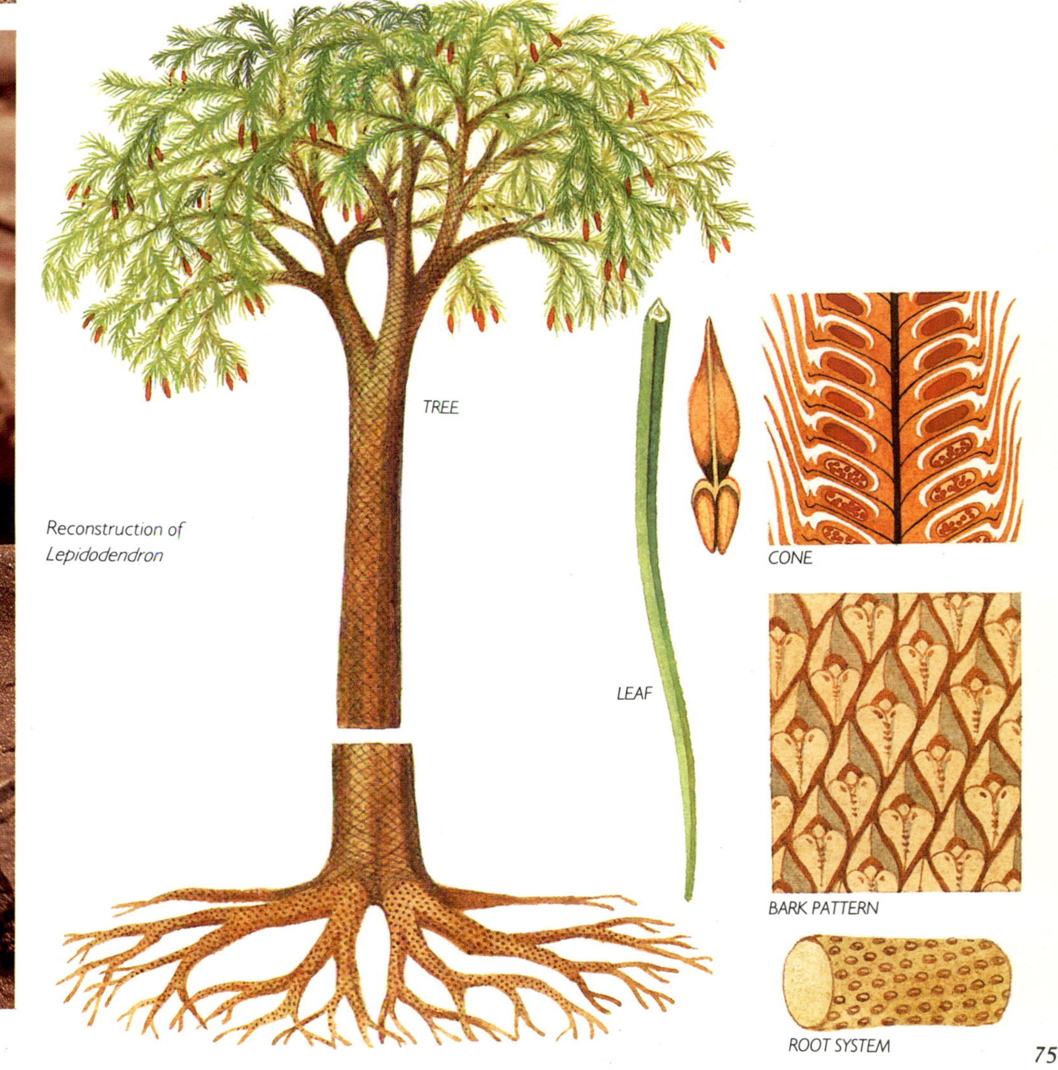

Reconstruction of *Lepidodendron*

82 *Lepidodendron veltheimianum: a branching stem with a pattern of inverted-drop shaped scars. From the Early Carboniferous Scartwater Formation in the Charters Towers region of Queensland, age approximately 340 million years. (× 6.0)*

as many as eight billion male spores per cone. The larger female spores would have numbered only hundreds per cone. Under the equatorial swamp conditions in which these plants grew in the Northern Hemisphere they formed dense forests, their immensely tall trunks only metres apart. There are petrified forests which show the trunks and root systems in growth positions, confirming the density of the forests.

Lepidodendron trees had root systems of thick, forking branches which were patterned with circular scars, each with a central spot representing the vascular trace to the rootlet. These root buttresses are known as *Stigmaria* and the rootlets which were attached to the circular scars were stigmarian rootlets.

The Giant Clubmosses grew only in swamps. When climatic changes resulted in the contraction of tropical swampy habitats with high rainfall, they ceased to dominate the vegetation. Their descendants have continued to exist in localised suitable habitats right up to the present day, though with a drastic decrease in size.

While *Lepidodendron* had cones as reproductive structures, other genera of Giant Clubmosses had other arrangements. In *Leptophloeum*, a common fossil of Australian Late Devonian strata, limited regions of young stems are fertile, bearing sporophylls instead of leaves. These zones have a distinctive pattern, different from the rhombic pattern of leaf-bases on mature stems.

37

Other genera known from Europe had produced very large female spores, retained them inside the sporophyte plant and effectively achieved a "seed".

While the *Lepidophyllum* leaves of *Lepidodendron* are long, narrow and substantial, *Leptophloeum australe* has small, inconspicuous leaves. They are like forking spines and are only very rarely preserved in fossils. Young stems bearing leaves have a leaf-base pattern of narrow rectangles, which later expands to the characteristic rhombic pattern of mature stems.

81 *A young stem of Leptophloeum australe, with leaves attached to rectangular leaf bases. This specimen was photographed in the British Museum of Natural History, London, and is one of the very rare examples with leaves attached to a stem. It was collected in the Broken River, North Queensland, in 1870.*

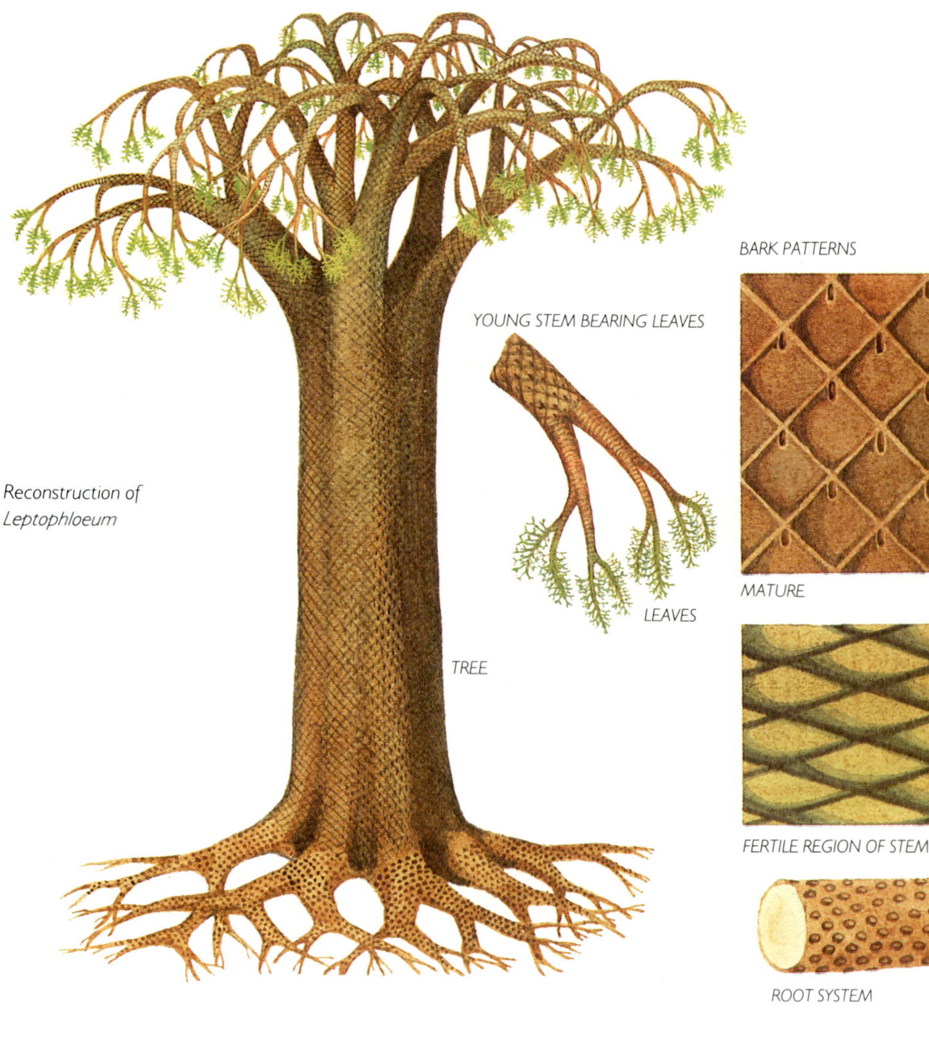

Reconstruction of Leptophloeum

83

85

84

86

LEPIDODENDRON

83 Bark pattern of Giant Clubmoss, Lepidodendron aculeatum, in which there is a mesh of raised fibrous tissue between the leaf base scars. Leaf base scars form an ascending spiral. From the Early Carboniferous at Charters Towers, Queensland, age approximately 340 million years. (× 4.4)

84 Lepidodendron mansfieldense, with a bark pattern of elongated lens-shaped leaf bases. This species ranges from the Late Devonian, approximately 370 million years ago, into the Early Carboniferous. This specimen is from the Drummond Basin, Queensland, from beds that are either Late Devonian or Early Carboniferous. (× 2.7)

85 A young stem of a Lepidodendron, with prominent leaf base scars which are almost circular. From the Early Carboniferous Scartwater Formation in the Drummond Basin, Queensland. Approximately 340 million years old. (× 3.1)

86 Lepidodendron canobianum, from the Cumberland Range Volcanics, Georgetown District, Queensland. Early Carboniferous age. (× 2.0)

87

LEPIDODENDRON AND *STIGMARIA*

87 A Lepidodendron stem with attached leaves. Young stems bore strap-like leaves in a bottle-brush arrangement. The leaves are the photosynthetic organs in this genus, in contrast to the arrangement in Leptophloeum, in which they were very small, and where photosynthesis must have been carried out by green twigs. This specimen from the Drummond Basin, Queensland, is approximately 340 million years old. (× 5.3)

88 Stigmaria ficoides, part of a root buttress of a Giant Clubmoss. The circular scars show attachment for stigmarian rootlets, and the central spot in each is from the vascular trace which supplied each rootlet. This specimen is from the Early Carboniferous in the Drummond Basin, Queensland. Age approximately 340 million years. (× 0.9)

89 This remarkable specimen shows a central stem with Lycopod bark markings, and the attached leaves form a continuous wing on either side. Presumably in life there were several wing-like lines of leaves, possibly giving a four-ranked arrangement, and the leaves appear to be thick and rigid, not like the flexible, ribbon-like leaves of other leafy Giant Clubmosses. From Telemon Formation at Emerald, Queensland. Approximately 355 million years old. (× 0.9)

88

89

90 *Lepidosigillaria linearis* (left) and *L. yalwalensis* (right), Lycopod stems with a pattern of leaf bases in vertical rows, forming a fine bark ornamentation. *L. linearis* is very similar to *L. whitei* of North America, and may not be a separate species. (Devonian floras were cosmopolitan.) This specimen is from Bunga Head, New South Wales, and is of Middle Devonian age, probably about 385 million years. (× 1.1)

40

90

91

LEPTOPHLOEUM AUSTRALE

91 *Leptophloeum australe*, a useful index fossil for the Late Devonian, common in Australia in beds of that age. The species may not be distinguishable from *L. rhombicum* of North America. It occurs also in South Africa, and is part of the cosmopolitan Devonian flora. The rhombic leaf-base pattern is characteristic of the species, and the vascular trace to each leaf is situated in the top angle of the rhomb. This specimen is from Barraba, New South Wales, and is approximately 370 million years old. (× 1.8)

92 Young, forking stem. The lower, unbranched portion, has the characteristic rhombic leaf base pattern, but the two branches have a rectangular pattern, and very small, spine-like leaves can be seen attached to some of the lower bases. This is an important specimen showing the connection between two different bark patterns and the attachment of leaves. Although the species is very common, only the characteristic rhombic pattern was identified until recently. It is specimens such as this that help us to reconstruct plants from fragmentary information. From the Late Devonian Telemon Formation, Emerald District, Queensland. Approximately 370 million years old. (× 2.8)

93 Very thin ultimate branch only 1.5 cm long and 1 mm wide, with double forked, spine-like leaves attached. This specimen is preserved close to the forking stem illustrated in picture 92, in the same rock. It is obvious from the very small nature of the leaves in *Leptophloeum* that photosynthesis would have been carried out by the thin stems and fine branches of the tree, which must have been green, and that the leaves played little part in the process.

94 *Leptophloeum australe.* The bark pattern in fertile regions of the stems was different from the characteristic regular rhombic pattern of mature stems. A pattern of interlocking horizontal diamond shapes, intermediate between the rectangles of leafy parts and the old bark pattern, indicates those regions of young branches which were fertile. The sporangiophores were borne in these areas instead of leaves. This arrangement of having a fertile region of a stem and not a well organised cone is a feature of the genus, and different from *Lepidodendron* which follows in the evolutionary sequence. The specimen is from Goonoo Goonoo, New South Wales, and is about 370 million years old. (× 4.3)

93

92

94

HORSETAILS OF THE GIANT CLUBMOSS FLORA

Reconstruction of an Arborescent Horsetail

Giant Horsetails, or Calamites, were co-dominants in the Northern Hemisphere swamp flora. They are represented in equivalent flora in Australia, but under local conditions they do not achieve the size and luxuriance of their northern counterparts.

Horsetails, ancient and modern, have jointed stems. They produce their leaves and branches in whorls at each joint and spores are produced in terminal cones. They have a very successful means of vegetative propagation, producing underground stems or rhizomes which run out from the main plant and put up erect branches to form new plants. The result is dense thickets like those of large Bamboos.

Some of the very large tree-forming fossil Horsetails were seed-producers, following the same evolutionary sequence of heterospory to retained female egg cell to seeds as had applied in the Giant Clubmosses.

When the ideal swamp conditions under which the Giant Horsetails achieved their greatest luxuriance disappeared, Horsetails of more moderate size continued to occupy suitable habitats and still do so today. They grew like rushes in swamps and along riverbanks throughout the ages, and are well represented in fossil floras because they grew where fossils are likely to be produced.

95

SEED-FERNS, FERNS AND EARLY GYMNOSPERMS OF THE GIANT CLUBMOSS FLORA

Fern-like foliage first appears in the Fossil Record in Late Devonian and Early Carboniferous times. It is not often possible to decide whether the plants which bore the leaves were Seed-ferns, Ferns or early Gymnosperms. Petrified woods show that all three plant groups were present. Occasionally fructifications are associated with leaves, clarifying their classification.

Some fructifications cannot be satisfactorily assigned to any existing plant group. *Barinophyton obscurum*, for example, may represent an evolutionary experiment. 41

96

96 *Adiantites paracasica, an Early Carboniferous Fern from Einasleigh in Queensland. The finely dissected foliage with forking pinnules is a very primitive arrangement. The species is known from South America. Age approximately 340 million years.*

97 *Barinophyton obscurum, a fertile early gymnosperm, is known only from a suite of fossils collected at a locality on the Genoa River in New South Wales. It appears to be indistinguishable from B. citrulliforme of North America, which is of the same Late Devonian age, approximately 370 million years. It may represent an evolutionary experiment, as its fertile branches with plate-like sporangia stacked one behind the other along a pinna axis are unlike anything known in other groups alive then or later.* 41

95 *Equisetum sp., a living Horsetail. Today only this one genus exists, distributed throughout the world except for Australia and New Zealand. All are herbaceous perennials; some die back each year to their underground rhizomes, other are evergreen. Leaves are small and fused into a leaf-sheath with only their points free. The growth habit with segmented stems and branches in whorls at nodes is clearly seen in this photograph. The Giant Horsetails in the Fossil Record had the same structural pattern, only the size was different. (Peter Valder)*

97

99

98

100

98 *Archaeopteris howitti.* A single leaflet and a feathery fertile branch with large numbers of elongated sporangia on the forking branchlets. This rare and important specimen comes from Iguana Creek, Victoria, from Late Devonian strata. Age approximately 370 million years. (× 1.4)

99 Enlargement of the fertile branch. (× 7.6)

100 *Archaeopteris howitti.* Small portion of a frond with Fern-like leaves, from Genoa River, New South Wales. It is very difficult to classify sterile foliage of this sort because the plant which bore it could have been an early Fern ancestor, or a Seed-fern, or a Ginkgophyte. The discovery of fertile material confirms the determination as *Archaeopteris*, a genus well known in Europe but until recently unconfirmed in Australia. (× 2.8) 41

101 This striking plant fossil from Howitt's Quarry at Bindaree, Victoria, is of Late Devonian age, about 370 million years. It appears to be a Lycopod with fleshy leaves but has not been identified. (× 2.3)

CHAPTER 4

THE *RHACOPTERIS* FLORA OF PRIMITIVE SEED-FERNS

THE MID TO LATE CARBONIFEROUS AND EARLY PERMIAN PERIODS

APPROXIMATELY 325 TO 285 MILLION YEARS AGO

By the Mid Carboniferous, Australia's rapid movement towards the South Pole and the onset of an ice age had had a marked effect on the vegetation. The Giant Clubmoss Flora died out. During the Late Carboniferous a flora of low diversity — the Rhacopteris *Flora of primitive Seed-ferns — survived the rigorous climatic conditions. About half of Australia was covered by a continental ice sheet at the climax of the ice age, which occurred at the end of the Carboniferous.*

After the Early Carboniferous, the opening out of the eastern side of Pangaea accelerated, sweeping Australia towards the South Pole. When the movement was completed, in the Late Carboniferous, the continent lay between 60° and 75°S and had rotated through 90°. It was to stay in high latitudes for 200 million years, until it severed its connections with Antarctica 50 million years ago.

With global cooling, and Australia and adjacent parts of Gondwana located in South Polar regions, the formation of an ice cap increasingly influenced the climate.

At first, glaciers appeared on the coastal volcanic highlands of eastern Australia. The highlands of the Centre were probably glaciated soon after. There is evidence that by the end of the Carboniferous a continental ice sheet similar to that which covers Antarctica today had spread northwards from the South Pole, and half of the Australian continent may have been glaciated. The maximum extent of the ice sheet occurred at the Carboniferous-Permian boundary in time. Marine transgressions in earliest Permian times reflect the melting of the ice sheet and the start of amelioration of the climate.

The Australian continental land surface had reached its complete size by 285 million years ago. The eastern region had been gradually added during the preceding geological periods. There were marine incursions during the Late Carboniferous and the Early Permian. Much of South Australia and Victoria were covered by shallow seas, and the Canning and Carnarvon basins in Western Australia were again inundated.

A Fern frond, showing the characteristic curled form of the unfurling new leaf. True Ferns have a long fossil history, back to about 325 million years, and similar foliage was borne by early Seed-ferns. (Jim Frazier)

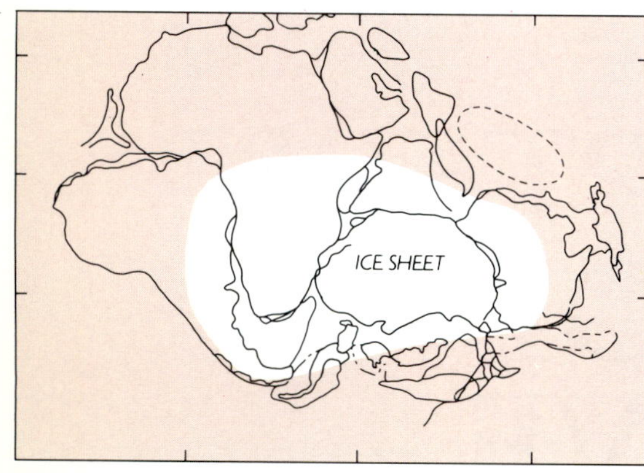

GONDWANA: EXTENT OF THE ICE SHEET.
Carboniferous-Permian boundary, 286 million years ago.

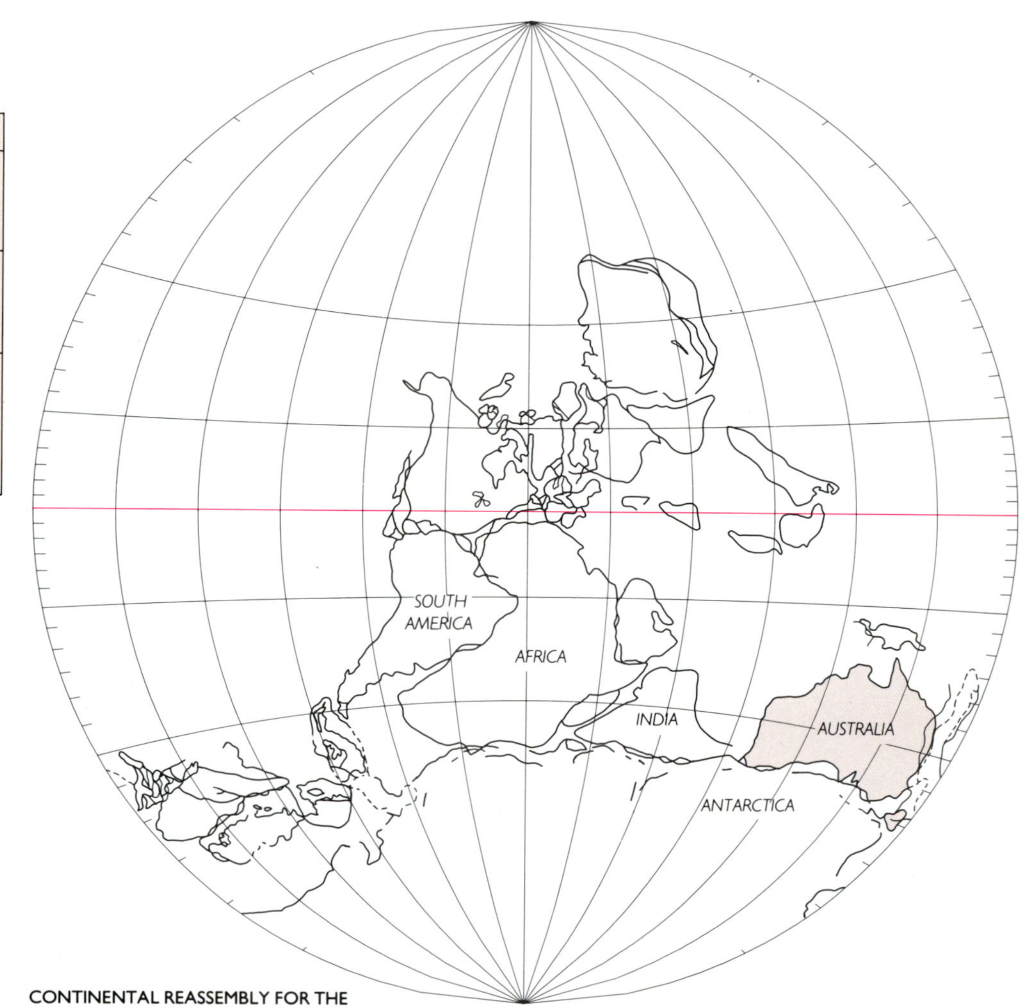

CONTINENTAL REASSEMBLY FOR THE CARBONIFEROUS-PERMIAN TIME BOUNDARY, *286 million years ago.*

102

102 Antarctic landscape. *(Alex Ritchie)*

The increasing cold of the approaching ice age had a marked effect on the flora. The already struggling Giant Clubmoss Flora of the Early Carboniferous, which had never managed to achieve luxuriance, disappeared.

Before the arrival of the next major flora — the *Rhacopteris* Flora, which is best developed in New South Wales — there is a gap in the Australian fossil sequence. This gap is largely due to major volcanic activity in the southern part of the Sydney Basin, which disturbed sedimentation. Some petrified woods are found, including *Pitys*, which is a gymnospermous wood from trees of considerable size. Early Gymnosperms probably grew during these times of alpine glaciation and marked seasonality of climate, just as their modern relatives do in sub-Arctic regions today.

The Australian *Rhacopteris* Flora is a poor relation of the Early Carboniferous flora of the Northern Hemisphere. Its impoverished nature is an expression of the extreme climatic conditions under which it grew. Its plants were able to adapt to the cold climate and have become slightly altered under the conditions of isolation imposed by distance. 42

The Australian *Rhacopteris* Flora is characteristic of the Late Carboniferous (whereas similar plants in the Northern Hemisphere are from the Early Carboniferous). Once again, the closest affinities of the Australian plants are with those of South America. The influence of climate is again the factor which determines the relationship.

The Rhacopterids were Seed-ferns with primitive leaf structure and venation. *Rhacopteris*, which gives its name to the whole flora, occurs throughout the Late Carboniferous sequence. By the middle of the Late Carboniferous, two other forms of Seed-fern appear. They represent opposite ends of a range of forms, from a type with large variously lobed pinnules to a type with deeply dissected pinnules, and in between them is a series of intermediate forms.

103 Rhacopteris ovata, the plant which gives its name to the Rhacopteris Flora. This Fern-like plant is in fact a Seed-fern. Its leaves are simple, with primitive venation — strong veins fanning out into the leaf blade. It probably grew like a ground Fern and certainly had to be robust and adaptable to persist throughout the Late Carboniferous time when the ice age made conditions harsh and unfavourable. When the cold became a limiting factor, and the Giant Clubmosses which were unsuited to changed conditions almost died out, Rhacopteris became the most common plant in a flora with very few species, in the Mid Carboniferous between 325 and 285 years ago. It is not in any fundamental way different from European examples, and its renaming as Pseudorhacopteris seems unwarranted. (× 1.8) 43

104

104 Botrychiopsis plantiana. A large, foliose example from Patterson, New South Wales. (× 0.9)

105 Dactylophyllum digitatum, a Seed-fern with very finely dissected pinnules, enters the Fossil Record in the second half of the Late Carboniferous at the same time as does Botrychiopsis. In D. digitatum, the pinnules fork into narrow segments. Associated with it, on the right hand side of the picture, is a leaf-like structure whose upper section is lobed irregularly. This structure is an aphlebium, believed to be a bract-like protective leaf which grew on the footstalk of the plant. Both Botrychiopsis and Dactylophyllum probably had aphlebiae. The specimen is from Stroud, New South Wales.

106 A Rhacopterid with partly divided leaves. A series exists from undivided pinnules to finely divided ones, and it is not easy to classify individual specimens. Leaves in the series are often referred to as the "Sphenopteridium Complex". This specimen is from the Late Carboniferous at Dingo Creek, Stroud, New South Wales. (× 2.6)

107 Botrychiopsis plantiana. This plant shows a lot of variation in leaf form. The frond on the left has small pinnules on the left side of the rachis and more foliose ones on the right, and the right hand frond has all large multi-lobed pinnae. The plant appears in the Fossil Record in the second half of the Late Carboniferous, and persists in rare instances into the Permian in Australia, and more commonly in India. Age about 305 million years. From Raymond Terrace, New South Wales. (× 1.2)

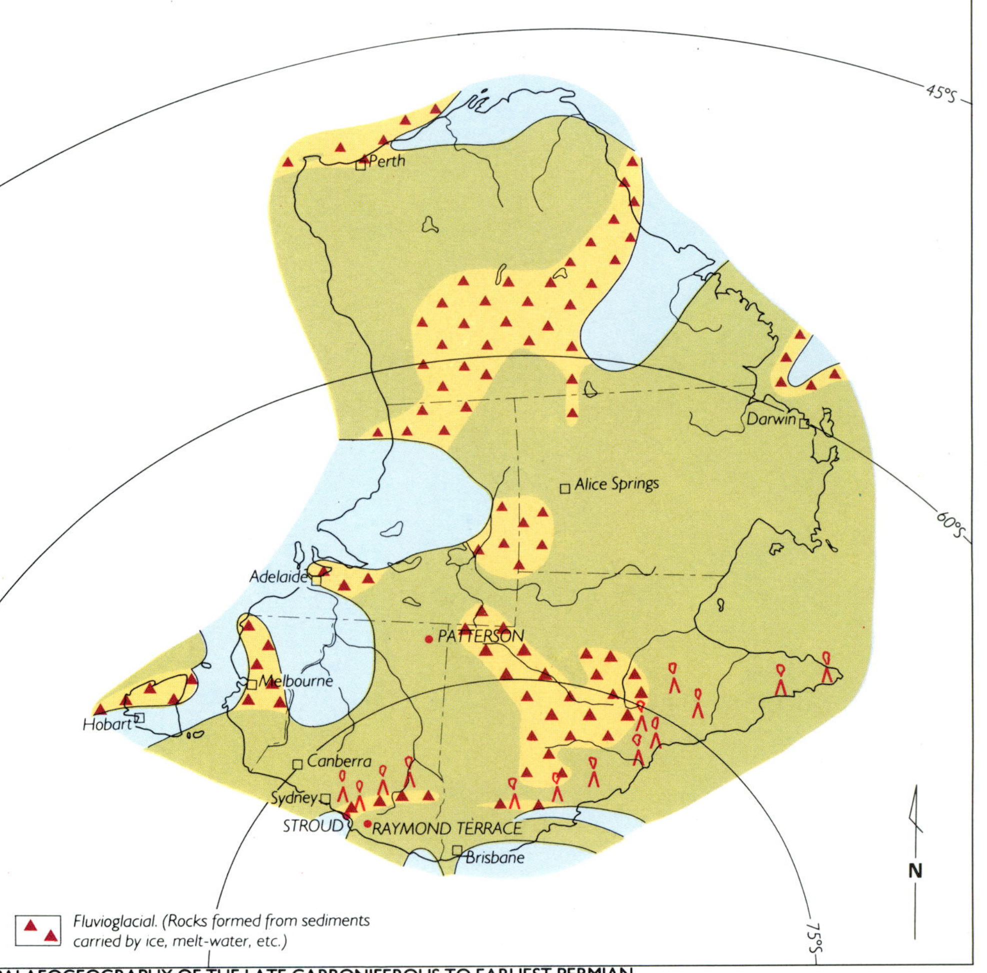

PALAEOGEOGRAPHY OF THE LATE CARBONIFEROUS TO EARLIEST PERMIAN.

There is some difficulty in naming this suite of plants. Those with foliose pinnae are *Botrychiopsis* and the very finely divided ones are *Dactylophyllum*, but the intermediates are less easily assigned to form-genera. Some authorities consider the intermediate forms to represent a hybrid swarm created by the interbreeding of the two distinct genera. These plants are also interesting because of their large aphlebiae, which are spathe or bract-like leaves that probably had a protective role. They may have ensheathed the more delicate fronds of pinnate leaves while they were developing.

43 44 45

Aphlebiae show a variety of forms, from big undivided leaves to ones in which part of the lamina is pinnate, approximating the normal foliage of the plant. Leaf-like aphlebiae are similar in size and form to the earliest Permian *Gangamopteris* leaves which appear in the interglacials. Venation of aphlebiae is simple, with no cross-connections between veins. It is possible that the Glossopterid leaf-type of the following Permian Period evolved from an ancestor with aphlebiae which lost the pinnate phase of its foliage under the extreme cold conditions at the height of the ice age. An evolutionary jump from primitive venation to netted venation is all that was required; and evolution of new characters is known to have proceeded in a series of jumps, not in gradual step-by-step fashion.

There were some Lycopods in the *Rhacopteris* Flora, derived from the Giant Clubmoss Flora and adapted to survive in the cold climate. Some of them show marked seasonal banding of their stems, due to times of growth alternating with seasons of dormancy.

The only member of the *Rhacopteris* Flora to persist into Permian times is *Botrychiopsis*, which is occasionally found associated with *Glossopteris*.

105

106

107

108

109

110

111

108 A Rhacopterid with intermediate type pinnules. From Stroud, New South Wales. (× 1.4)

109 A large, almost entire aphlebium. When originally described, such leaf-like structures were given the name Rhacophyllum diversiforme. (× 0.9)

110 A group of fruiting bodies of Dactylophyllum (in the top central part of the picture).

111 Rhacopterids with finely dissected pinnules referable to Dactylophyllum, intermediate type pinnules and an aphlebium which is strongly pinnate are seen in this specimen from Stroud, New South Wales. (× 0.9)

ICE AGE LYCOPODS

112 A Lycopod stem, Subsigillaria sp., which shows marked seasonal banding. Two periods of dormancy are indicated by the two zones where the leaf bases are undeveloped. The growth season leaf bases in the wider middle section of the stem show gradual increase in size as the season advanced, then uniform mature bases towards the top, followed by the abrupt line of the dormant zone. These Lycopods probably were low, shrubby plants, growing in swampy areas and adapted to survive during the very cold winters when their watery habitats probably iced over. The Rhacopterids grew at the same time, forming a low-growing hardy vegetation like that of alpine habitats today. This specimen is from Stroud, New South Wales. Age approximately 305 million years. (× 1.8)

113 Cyclostigma australe. This Clubmoss has a pattern of small, circular scars on its stem. Fossils of this plant always have a glossy patina as though the surface was cutinised. It may be that as the plants grew under cold climatic conditions, they had a thick protective film. From Smith's Creek, Stroud, New South Wales. Late Carboniferous age, about 305 million years. (× 1.4)

114 A fertile organ, of unknown affinities, which is associated with pinnules of Dactylophyllum in a specimen from Clarencetown, New South Wales. Late Carboniferous age. (× 3.9)

112

113

114

CHAPTER 5

THE *GLOSSOPTERIS* FLORA

THE PERMIAN PERIOD

APPROXIMATELY 275 TO 245 MILLION YEARS AGO

After the ice age of the Late Carboniferous and earliest Permian times, the climate warmed up and there was rapid evolution of a rich flora characterised by Glossopteris *plants. Extensive cool-temperate swamps with thriving plant communities formed coal deposits. Early Ginkgos and Conifers appear in the Fossil Record. They colonised the drier hillsides and were not restricted to swamps. Tree-ferns, similar to those alive today, were abundant.*

Ginkgo biloba, the Maidenhair Tree, the modern representative of an Order of Gymnosperms which has a long fossil record. Ginkgophytes start to appear in the Permian in Australia. *(Jim Frazier)*

The contracting of the continental ice sheet in earliest Permian times was accompanied by a rise in sea level as the ice melted. As a result, large parts of the Australian land surface were flooded.

The seas retreated first from the southern part of the continent. The Canning and Carnarvon basins of Western Australia and most of the Bowen Basin in Queensland remained flooded by shallow seas until the middle of the Period. Then folding and uplift in the New England region resulted in a decrease in the area submerged on the margin of the continent, restricting it to an embayment lying between Rockhampton in Queensland and Grafton in New South Wales.

The climate warmed up gradually after the ice age. Nevertheless, it was a cold climate for most of the Permian with alpine glaciers on higher ground, particularly on the mountains of the south-eastern quarter where much high ground had been generated during the Kanimblan Orogeny of the previous Period. Rock sequences from sediments accumulated in the Sydney Basin in Permian times have ice-rafted boulders and dropstones in them, providing evidence of freezing winters. Seasonality of climate is also shown in the marked growth rings in the wood of trees petrified during this Period.

The climate in northern Queensland, Western Australia and the Top End of the Northern Territory may have been somewhat warmer and, though strongly seasonal, summers were probably temperate rather than cold-temperate.

Coal swamps were features of the landscapes, occupying large areas in drainage basins, particularly in the eastern half of the continent. Smaller coal swamps occurred in Western Australia and in South Australia.

The marine transgression in the North-West Shelf area of Western Australia during Permian times was the forerunner of the delta formations of the Triassic as the seas retreated. The sediments which were to accumulate were the main reservoirs for the gas fields of the area. The gas was in part derived from marine organic matter incorporated in the earlier sediments, but the main derivation was from the plant remains in the terrestrial sediments from the Triassic deltas.

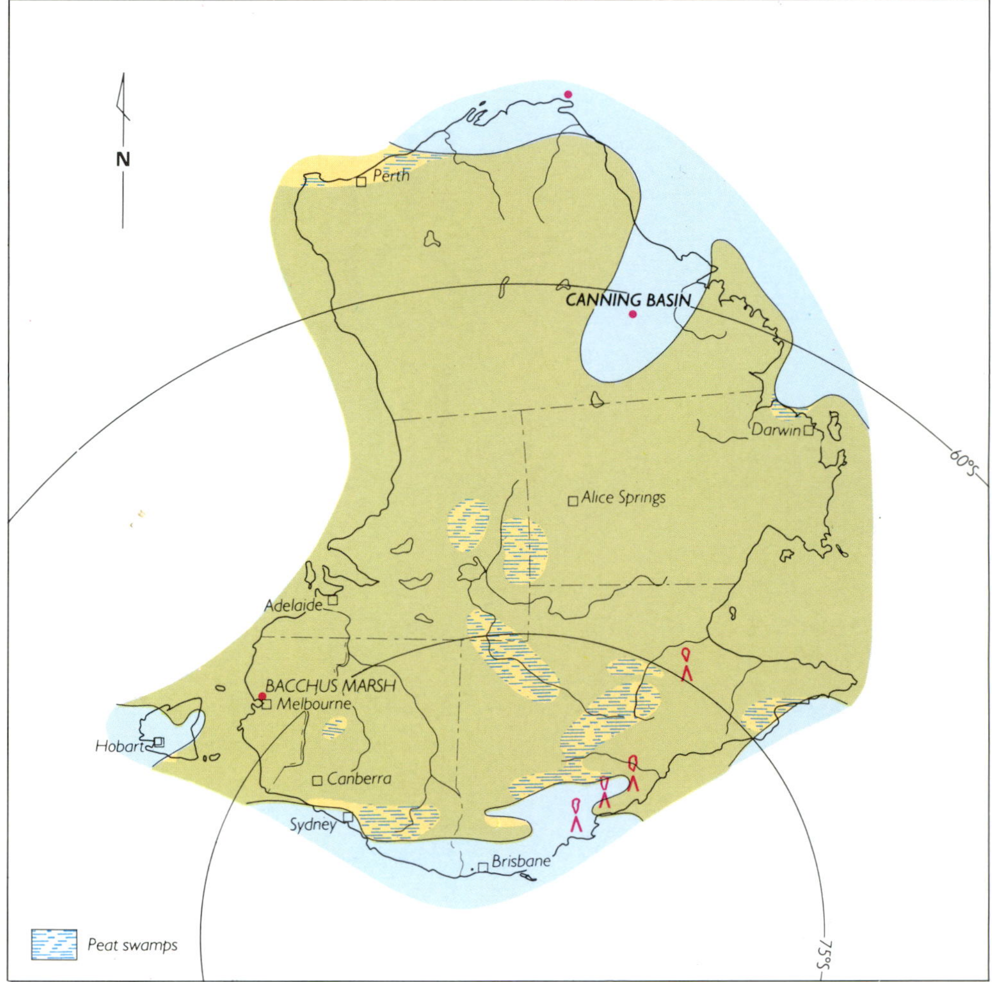

PALAEOGEOGRAPHY OF THE EARLY PERMIAN.

115 *A 50 year old tree with a trunk only 6 cm in diameter! This segment of petrified tree trunk showing very close annual rings, is from the Werrie Basin, New South Wales. Its age is Early Permian, about 275 million years. The very slow growth and uniformly small annual rings indicate a cold climate with dormant winters and only short growing seasons, which are in keeping with the analysis of the climate of the times. Australia was then between 45° and 90°S. The ice age conditions were still being felt, with glaciers advancing and retreating, and tree growth was stunted as in alpine areas today. (× 1.2)*

115

PALAEOGEOGRAPHY OF THE LATE PERMIAN.

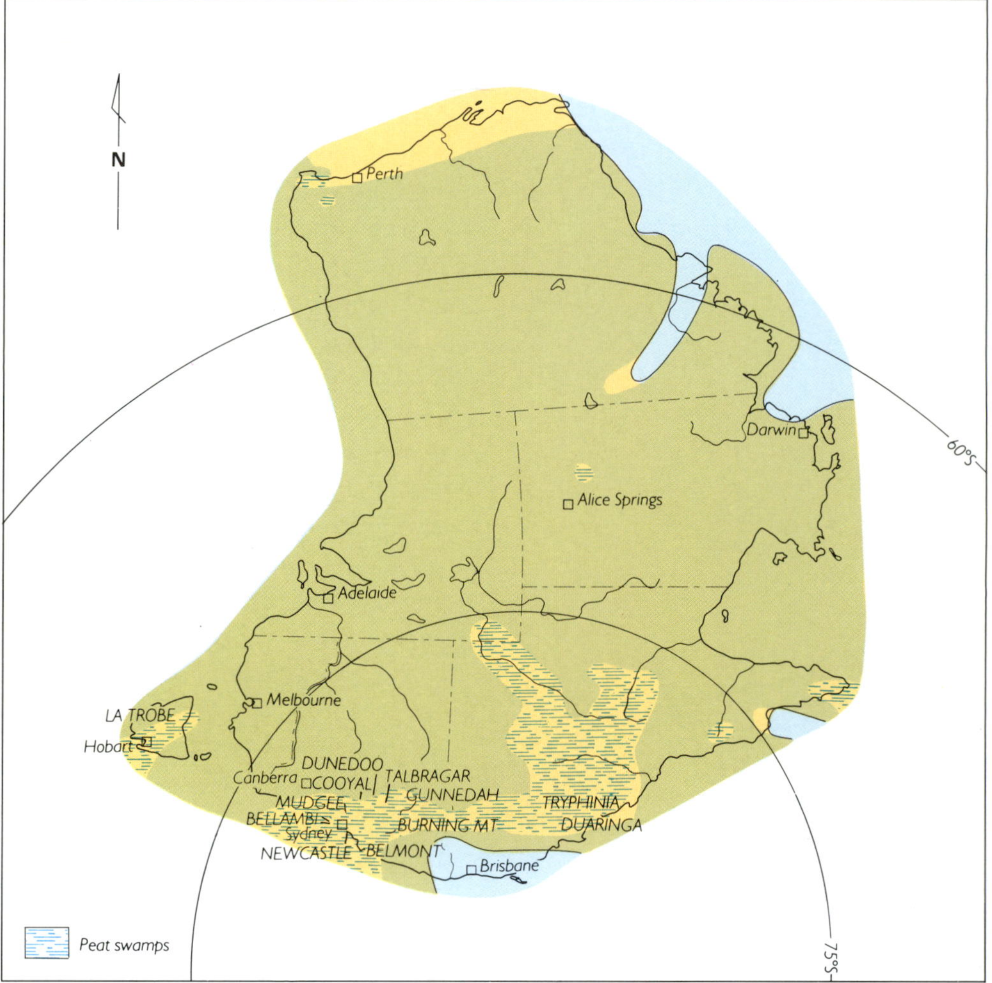

116

116 Palaeosmunda williamsi, a petrified Tree-fern trunk of the Late Permian, from Blackwater in Queensland, age about 255 million years. Related modern Tree-ferns in OSMUNDACEAE *have similar stem structure. Tree-fern trunks are complicated compound structures of main stem, frond bases, roots, scales and fibrous material. This specimen shows a double trunk with two main stem vascular cores. (× 1.6)* 79

PERMIAN SEDIMENTARY BASINS.

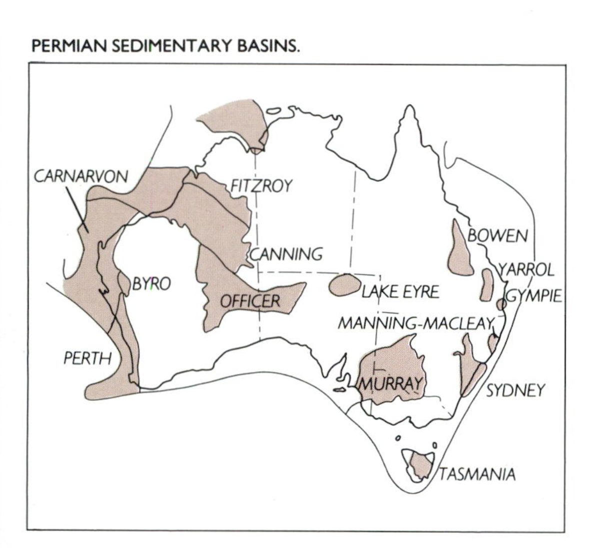

The improving climate prompted a great explosion of plant life. The thriving *Glossopteris* Flora evolved from the floral remnants which had survived the rigours of the ice age.

The immense coal deposits laid down during the Permian are being exploited today and are an important part of the Australian economy. The stereotype of coal swamps being steamy tropical swamp-jungles, as was the case in the Carboniferous in the Northern Hemisphere, is not applicable to southern Permian coals. These coal deposits are the product of cold, swampy bogs in which Horsetails grew in immense profusion like rushes. Ferns and Seed-ferns, and probably Mosses like modern peat, as well as herbaceous Lycopods like *Selaginella* formed a dense, low swamp vegetation. Trees and shrubs of the Glossopterids, with special aeration tissues in their roots suited to the boggy conditions, grew in the swamps and in adjacent areas with high water-tables.

By Late Permian times the drier hillsides and places away from permanent water were habitats for early Conifers, Ginkgos and Cycad-ancestors. This evolution of a vegetation adapted to living in drier places away from the swamps was a major advance. Efficient root systems, leaves adapted for water conservation, and reproduction by seeds enabled the transition. Symbiosis between roots and Mycorrhizal Fungi was probably a universal feature.

Because of the severe climate with very cold winters, the Glossopterids were deciduous. Ginkgos probably also shed their leaves annually. Other plants which retained their leaves would have experienced months of dormancy followed by seasons of active growth.

THE GLOSSOPTERIDS

46-69
74-76

Glossopteris leaves, the common and widespread fossils of Permian strata in southern lands which were once united in Gondwana, are still among the least satisfactory plant fossils in terms of their classification and nomenclature. This problem has arisen because of the wide variety of venation patterns they present, the difficulties in delineating specific characters and the problems inherent in assigning to "species" on leaf morphology alone.

The name *Glossopteris* is derived from the Greek words *glosso*, meaning "tongue", and *pteris* meaning "fern". The famous European systematic botanist Adolph-Theodore Brongniart created the name in 1822 for a tongue-shaped "fern" leaf which he described, only to find later that the organ which he had described was in fact a *Lepidophyllum* (a leaf of a *Lepidodendron*). The name was used in 1830 for a Permian leaf from India with net venation, pronounced midrib and a tongue shape — a true *Glossopteris* as we now define the form-genus. Many "species" were described during the last century and illustrated by artists' drawings, some of which were inaccurate and "touched up" for artistic reasons. The careful, wordy descriptions of the first described "species" and the drawings of type specimens have been used for naming of specimens ever since. Palaeobotanists have tried dutifully to match new specimens to the original descriptions as if they were gospel. Inability to locate and re-examine most of the specimens described in the pioneering work of Dr Ottokar Feistmantel, from as early as 1876 in collections in India (where they have been lost, or may be hidden in uncurated collections), is another problem.

46

117 A magnificent specimen of *Glossopteris* leaves on display in the Australian Museum, Sydney, showing leaves of different sizes, and one leaf of *Noeggerathiopsis hislopi*, (top of picture, right of middle). From Newcastle, New South Wales. Late Permian age, 255 million years.

118 *Gangamopteris angustifolia*, one of the first Glossopterids in the Fossil Record. From Bacchus Marsh in Victoria, where it occurs in glacigene sediments. Age Early Permian, about 280 million years. The leaf has a median groove, no defined midrib. Because of the coarse and relatively unconsolidated nature of the sediments, no detail is visible. (× 1.6)

118

The advent of good photography, high quality printing and good communications has helped to clarify the situation somewhat. However, deciding what name to give to each of the great number of detached leaves known only as impression fossils is largely an academic waste of time and tells us little of the evolutionary status or nature of the plants which bore the leaves. What is of interest is that leaf-forms in the Glossopterids were diverse. By the end of Permian times some leaves had a midrib and a network of secondary veins like Dicotyledons; others had parallel veins like Monocotyledons and some Conifers; some were taeniopteroid like Bennettitalean Cycads; and some were very small, probably growing on plants which were woody herbs. All the leaf types were here in this Order of plants, providing a genetic reservoir for the evolution of plant groups which arose in the Mesozoic Era and which may in some instances be on a direct line to modern plants.

The essential character in Glossopterid leaves is the presence of cross-connections between lateral veins, resulting in a mesh or net-venation pattern on the leaf lamina. When there is a midrib, the leaf is classified as *Glossopteris*. When there is a median groove in the leaf but no midrib, and there is also a pattern of meshes on the lamina as in *Glossopteris*, the form-genus name of *Gangamopteris* is given. When there is no midrib and only few cross-connections between veins on the lamina, *Palaeovittaria* is the name chosen.

In *Glossopteris* the midrib is not a single mid-vein but is a stranded structure of a number of veins. When they lie close together they appear as a single strand, when more widely spaced the multiple strands are clearly seen. Leaves are often grooved along the median line and some have a keel of thickened tissue giving extra support. There is a graduation from leaf types in which the midrib is clearly defined (classified as *Glossopteris*) to types in which the strands are well spaced and not aggregated into a midrib (classified as *Gangamopteris*), and it can be difficult to decide whether there is a midrib or just a median groove when preservation of the fossil is poor. A similar gradation exists between leaves classified as *Gangamopteris* and those at the other end of the spectrum known as *Palaeovittaria*.

119 Gangamopteris walkomii. The growth habit of this plant is clearly shown in this specimen from Narrowneck at Katoomba, New South Wales. Venation of leaves shows that the criterion of a midvein being present or absent is hard to apply even in this one twig. About 255 million years old. (× 1.6) 76

120 Palaeovittaria sp. A Glossopterid in which there is very little meshing of veins and no median groove as in *Gangamopteris*, but where the relationships of the plant are known because this sort of leaf has been found bearing reproductive structures of the same sort, and in the same manner, as in Glossopterids. Specimen from the Insect Beds, Belmont, New South Wales. Late Permian age. (× 1.0)

121 Gangamopteris cyclopteroides from Bacchus Marsh, Victoria, where it represents the first Glossopterid to appear in the Fossil Record, appearing in the sediments produced by the glaciers of the Late Carboniferous to Earliest Permian ice age. Age about 280 million years. (× 0.9)

122 Cuticle of a *Glossopteris* leaf. This photograph of a microscope slide preparation of a cuticle from the Richmond Vale Colliery, New South Wales, shows the characteristic venation clearly. The stranded midrib of several veins, and the meshing of lateral veins is diagnostic of the genus.

123 *Glossopteris* leaf, showing a prominent midrib and a mesh of lateral veins, with a leaf of *Noeggerathiopsis hislopi* lying across it. The strong, parallel venation of the Cordaite is clearly seen. Specimen from Newcastle, New South Wales. (× 2.5)

119

120

121

122
123

124

125

All the leaf form-genera of the Glossopterids are known to bear similar fructifications, so they are legitimately classified together into one Order. Classification of living plants depends on a full understanding of reproductive structures of individuals and their grouping into Orders, families and genera according to the nature of their fertile organs. No satisfactory classification of modern plants could be made on leaf morphology. However, where fossil plants are concerned and only unattached leaves are known, each leaf type has to be given some name as a descriptive identification. This "classification" of detached organs was much emphasised in the past and is gradually being replaced by a scientific classification comparable to that used in modern plants, as our knowledge increases and whole plants can be reconstructed from the fragmental fossil evidence.

The Glossopterids in Gondwana were comparable to the Eucalypts in modern Australia in that they dominated the vegetation of the whole landmass in the way that Eucalypts dominate the vegetation of the continent today. But the Glossopterids were a complete Order of plants, and the similarity of their leaves and their classification into *Glossopteris* and a few other genera is misleading. *Eucalyptus* is a genus (though it may shortly be revised as several closely related genera) with 450 species. Some genera of the same family (MYRTACEAE) as the Eucalypts have similar leaves but different floral anatomy. Since the floral anatomy of all the species is known, their classification is scientific. There is no need for subjective judgements of whether sizes and angles of meshes in the lateral venation of the leaves are constant or different, or whether a leaf is juvenile or fully expanded, or other considerations of that sort.

Only rarely are Glossopterid leaves found attached to stems. The vast numbers of leaves preserved in layers in some shales are due to the deciduous habit of Glossopterid trees and the accumulation of "autumnal banks" in lakes and estuaries when leaves were shed.

Glossopterid leaves come in all sizes, from *Glossopteris ampla* which has "fronds" up to a metre in length and almost as wide, to more moderate sizes and many very small leaves. Leaf shapes vary from short and broad to long and narrow, with tongue-shaped, lanceolate, oval or any other shape between. Leaf apices may be blunt, pointed or indented. The blade may taper towards the stem or bulge out into a heart-shaped base, or it may even have swallow tails as in *Glossopteris duocaudata*. 47
Leaves may have long or short petioles (leaf stalks) or none. Venation may be fine and regular or it may have larger meshes along the midrib or a pattern of large meshes all over. The angles of lateral veins with midrib and margin varies, some sweeping upward, others curving up and arching out to meet the margin at right angles, and still others running out straight at right angles to the midrib and parallel to each other. There is a bewildering variety.

The few specimens which have been found with attached leaves show that the leaves were borne in whorls, or close spiral arrangements, at the end of small branches. A pattern of scars on the branches shows where the previous season's leaves were attached.

The Glossopterids were woody plants — trees and probably shrubs — presumably of all sizes, suiting them to different habitats and niches. Their wood anatomy is of *Araucarioxylon* type (a gymnospermous wood, with regular sized tracheids). 48 49 50

The root systems of the Glossopterids were adapted to their swampy habitats. They had a specialised internal structure which probably had an aerating function. These roots are known as *Vertebraria* because, with their segmented cores, they look like vertebral columns or back-bones. 51 52

Some specimens show *Vertebraria* in continuity with trunks with Glossopterid wood structure and, as trunks have also been found with branches giving rise to

smaller branches that bear leaves, the vegetative parts of the plant are known to belong together. In a few instances, by the arrangement of branches and leaves, we can form an idea of how some of the trees looked when alive. We can only make an educated guess about the shrubby forms and the woody herbs which were part of the spectrum of types by the Late Permian. There is no reason why they should have looked "prehistoric". Their general appearance must have been like leafy trees and shrubs and woody herbs today.

The first Glossopterid leaves to appear in the Fossil Record are *Gangamopteris cyclopteroides* and *Gangamopteris angustifolia*, found in the glacigene sedimentary rocks at Bacchus Marsh in Victoria. The sediments forming these rocks were produced by the activity of glaciers during the ice age of the Late Carboniferous and Early Permian times. In the Sydney Basin the first Glossopterids appear in the Dalwood Group of the Lochinvar Formation of the Lower Marine Series and are of Early Permian age.

Although the origin of the Glossopterids is highly speculative it seems possible that their ancestry might be found in an aphlebiate plant of the Carboniferous which lost the pinnate phase of its foliage.

127

126

126 A small Glossopteris leaf, from the Insect Beds at Warners Bay, New South Wales. (× 5.4)

127 Three different venation patterns are seen in this Glossopteris specimen from the Late Permian Illawarra Coal Measures, at Cooyal, New South Wales.

VERTEBRARIA INDICA

124 Vertebraria indica. From Newcastle, New South Wales. The name given to these root fossils reflects their similarity in appearance to the vertebral columns or backbones of animals. (× 0.5)

125 Vertebraria indica. From the Balmain Colliery, Sydney Harbour, New South Wales. The segmented core of the roots of a Glossopterid plant are an adaptation to a swamp habitat and presumably had an aerating function. (× 0.5)

128

129

130

128 *Glossopteris* leaves and a fragment of Fern from the Late Permian Illawarra Coal Measures at Dunedoo, New South Wales. The plant impressions are stained with iron pigments, making the specimens dramatic and much sought after by collectors, and by builders who use the shale with its plants for feature walls in buildings. (× 2.0)

129 *Glossopteris duocaudata*, the leaf at the left of the picture, has two swallow-tail extensions at the base of the leaf blade. From Cooyal, New South Wales, in Late Permian, Illawarra Coal Measures. About 255 million years old. (× 1.1)

130 *Glossopteris* leaf with an indented apex. From Dunedoo, New South Wales. (× 3.3)

131 *Gangamopteris walkomii*. This twig of the plant shows leaves borne in close spiral or whorled arrangement, and groups of small scales clustered in the centre at the top of the axis. Whether or not these are bud scales to protect the growing tip and to act as an over-wintering bud when the leaves fall, if this plant is deciduous like many of its relations, is not clear. From Narrowneck, near Katoomba, New South Wales. Of Late Permian age. (× 1.7)

132 *Glossopteris linearis*. A woody stem, seen in impression (lower left) and cast (above, left) with a pattern of leaf scars, from which leaves of previous seasons have been shed. Long, narrow leaves in close spiral arrangement are attached at the tip of the stem. Part of the stem was petrified, and sections cut and examined under a microscope showed that the wood is of *Araucarioxylon* type, like the wood of modern Hoop and Norfolk Pines. This is an important specimen because it shows organic connection between leaves and a definite wood type. Specimen from Mudgee, New South Wales. From Late Permian, 255 million years old. (× 1.0)

131

132

GLOSSOPTERIS VENATION PATTERNS

133

134

135

136

138

138 *Plumsteadia ampla.* An example of a large, multi-ovular fruit with its seeds not obscured by a cover-leaf. From Duaringa, Queensland, about 255 million years old. (× 0.9)

REPRODUCTION IN GLOSSOPTERIDS

Glossopterid leaves had been described and speculated about since early last century, but in the absence of fertile material their classification was uncertain. It was not until 1952 that reproductive structures were found attached to leaves. Dr Edna Plumstead in South Africa described a number of seed-bearing fruits attached to different leaves in publications in 1952, 1956 and 1958, and a new era started in the study of the Glossopterids. 54 a–e

Since then, the understanding of the reproductive biology of this group of plants has been ever increasing. There has been much work in India and in Australia, and although much has still to be learnt, a picture is emerging. It is apparent that the similarity of leaves is not an indication of uniformity in reproductive mechanisms. As knowledge increases and more examples are found of different fertile structures attached to leaves of particular "specific" types, a new classification will be possible. And this time the classification will be a scientific one based on reproductive structures. Then the form-genera names *Glossopteris*, *Gangamopteris* and *Palaeovittaria* will be useful for describing Glossopterid leaves whose true affinities are still unknown because they have not been found with attached fertile structures or are not themselves attached to stems which also bear fertile organs. The new genera

RECONSTRUCTION OF A *GLOSSOPTERIS* TREE

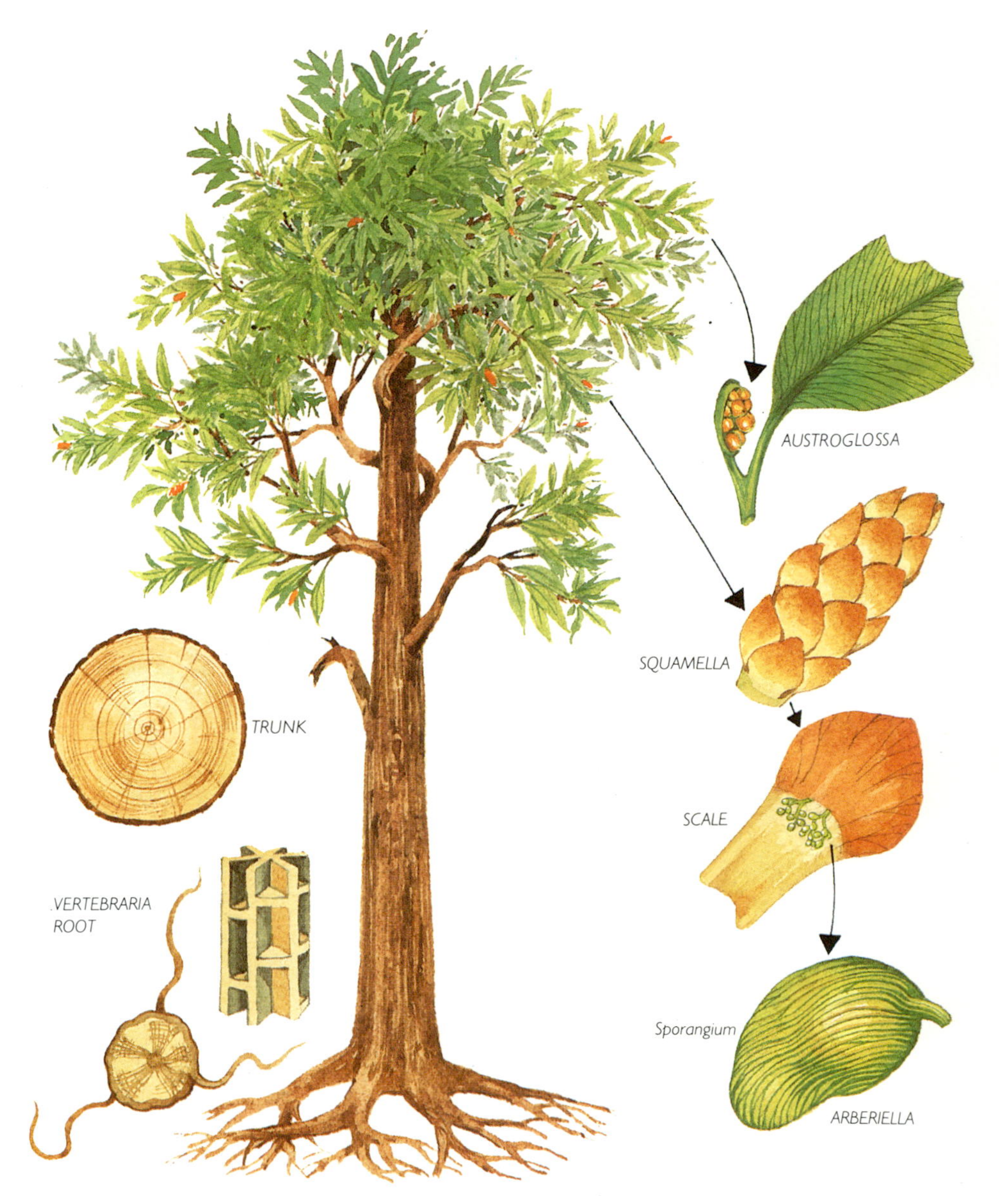

GLOSSOPTERIDALES —
RECONSTRUCTION OF MEGAFRUCTI

then described will be the names by which each whole plant unit is identified, as in modern plants. It seems logical that the choice of name should be that of the female fruit type as this has proved to be the character which varies fundamentally in different sections of this plant group.

FEMALE REPRODUCTIVE STRUCTURES IN GLOSSOPTERIDS

The female reproductive types of Glossopterids are significant in the evolutionary story. The different types which evolved are precursors of the fruiting bodies that characterise several groups of plants in the Mesozoic Era, whose origins might well be in this gene pool.

There are two distinct types of female structure, and they are so obviously different that they divide the Glossopteridales into two Sections.

In the first Section the "fruits" are massive, have many seeds and are carried on unmodified foliage leaves. The evolutionary sequence within this group involves reduction in the number of seeds in the fruits, and there is also an evolutionary trend towards the fusion of the protective cover-leaf or bract, which is part of the fruit, with the seed-bearing receptacle. This Section, for purposes of inclusion in a new classification, can be designated *Glossopteridales*, *Sect. Megafructi*.

In the second Section the seeds are attached to modified leaves or scale-leaves. (A similar situation exists in male fertile structures known to belong to Glossopterids, where the sporangia are borne on modified leaves or scale-leaves.) Classification of this Section is *Glossopteridales*, *Sect. Microfructi*.

SECTION MEGAFRUCTI

The first attached reproductive structures, described by Dr Plumstead in South Africa, all belong to this Section. They were greeted by the botanical community with interest and speculation began about their significance in the evolutionary context, especially as possible "missing links" to the Flowering Plants. Because they occurred as impressions or structureless compressions and many were poorly preserved, details of their structure remained a mystery. Since then, many examples have been found in India and in Australia. As a result, the structure of some of the different types is better understood and it is possible to make reconstructions. 55a b 56

Essentially, all the Megafructi types are borne on normal leaves. They are either stalked or sessile, attached to the petiole or to the blade of the leaf anywhere along the midrib. How and where they are attached has enabled the separation of a number of "genera", which have been described. Once again, names given before there was complete understanding of the fruit types were often premature. Some of them relate to differences created by methods of fossilisation or deformation during the process, or to the state of maturity of the organ, rather than to diagnostic features. 57a

To avoid confusion the names used in this book are "descriptive" and not "generic", and this must surely remain the course to take until a new classification based on reliable criteria is possible.

The basic arrangement appears to be "fruit" composed of a receptacle (or core) with seeds attached all around it, and a cover-leaf (or bract) which gives it protection while developing and probably falls off when the seeds are ripe. The receptacle was either cylindrical (or oval in cross-section by flattening dorso-ventrally) or spherical, and the appearance of all the seeds is uniform.

From the basic-type fruit with a receptacle with seeds attached all around it and a free cover-leaf, there are several lines of evolution. Some involve fusion of

139 A Plumsteadia fruit, part broken away, with its cover-leaf seen behind it (right) at the base of a Glossopteris leaf, in a specimen from Late Permian at Mudgee, New South Wales. Age about 255 million years. The seeds, which seem to be immature in this specimen, are on the outside of a receptacle which is conical. (× 5.5)

140 Plumsteadia ampla. A large, mature fruit with seeds seen to be surrounding the receptacle (second layer, below). (× 1.6)

141 Plumsteadia sp., an elongated structure with many seeds. This specimen comes from Mudgee, New South Wales.

142 Plumsteadia ampla, a large, probably fleshy, fruit which is strongly cycadean in appearance. It has a stout stalk, and a spherical receptacle with seeds attached all around it. The seeds are only partially seen in this specimen, as they are below the cover-leaf (the fruit being seen from that angle in this case). Where the fossil is broken to the left of the stalk, seeds can be seen exposed (cover-leaf has been removed). This specimen is from Duaringa in Queensland. Late Permian age, 255 million years. 56–57

parts of the fruit (with each other or with the foliage leaf on which they were borne), some show a reduction in the number of seeds, and in others there is a combination of both tendencies.

Plumsteadia, *Scutum* and *Dictyopteridium* have the fruit, stalked or sessile, attached to the leaf base or the petiole of the foliage leaf. Seeds are either wingless (or with a narrow wing all around) or have a pronounced wing (*Indocarpus* type). *Plumsteadia* and *Dictyopteridium* have the former type; *Scutum* has the latter, resulting in an appearance of a fluted ring around the compressed fruit.

In the specimens described as "*Lanceolatus*" in South Africa, the sessile fruit is attached further up the leaf and may even be fused to it, particularly in the young stages. These specimens, examined recently in the Plumstead collection in Johannesburg, were too poorly preserved for any opinion to be formed as to whether the cover-leaf was fused to the seed-bearing organ or not. The fruit appears to be the same general type as *Plumsteadia*, with uniform round seeds, but the receptacle is much flatter and probably unifacial.

140

141

142

143

DICTYOPTERIDIUM SPORIFERUM

143 The fruiting body in this specimen (top) shows the full structure of Dictyopteridium sporiferum. In the centre is a receptacle, ornamented with small spots, and in a fringe on either side are the very thin, tall seeds still attached to the receptacle. Specimen from Newcastle, New South Wales. (× 1.4)

144 Dictyopteridium sporiferum. The oval organ with a pattern of small spots on its surface is an example of the most common form in which this fruit type occurs. It represents the receptacle. The seeds have fallen off, and the spots are the points at which they were attached. Specimen from Cooyal, New South Wales. Late Permian age, 255 million years. (× 1.3)

145 Scutum sp. from La Trobe, Tasmania. The seeds borne on the receptacle in this organ are Indocarpus type with a wing, and as a result when the structure is compressed it appears to have a fluted margin. Age 260 million years. (× 6.6)

PAGE 114

146,147 A new Glossopterid fruit from the Insect Beds at Belmont, New South Wales. Late Permian age, 255 million years. The receptacle with its attached seeds is held between two cover-leaves. In one half of the specimen the fruit is seen with a cover-leaf behind it; and in the counterpart a separate cover-leaf with a lobed margin is preserved. (× 4.7) 65

148 Austroglossa walkomii. A fructification consisting of a small bunch of narrow-winged seeds is attached to the petiole of a Glossopteris leaf by a short pedicel. This specimen is from Cooyal, New South Wales. From the Late Permian, about 255 million years old. There is an evolutionary tendency towards reduction of the number of seeds comprising fruiting bodies in the Late Permian Glossopterids, and Austroglossa represents a stage in the trend. (× 4.1) 59

An Australian fertile structure described as *Senotheca* carries the fusion to the leaf a stage further, apparently consisting of two rows of seeds attached on the edges of the midrib.

Two recently described genera from India are *Jambadostrobus*, in which *Plumsteadia*-type fruits are attached two or three to a leaf on the midrib in the mid region of the lamina, and *Venustostrobus*, in which a *Scutum*-type fruit with a wing and a cover-leaf is attached in the middle of the leaf. 58

Another type of fruit, described as "*Ottokaria*" by Plumstead, has a long stem and the head appears to be flattened, with a prominent wing and a free cover-leaf. *Ottokaria* from India, where the genus was originally described and where recent research has enabled a reconstruction to be made, is a *Plumsteadia*-type fruit in which the cover-leaf is changed into a conical wrap-around structure. The generic name should be restricted to the first organ to which it was applied and to interpretations of that organ. In this case it is the Indian specimens and reconstruction which define the genus. Plumstead's examples should be included in *Scutum*. 54 b 55

Reduction of the number of seeds in Megafructi is seen in *Austroglossa*, where the fruit has become a small number of platyspermic seeds of *Nummulospermum* type attached to a little branch which is joined to the petiole of a *Glossopteris* leaf. 59 60 a

The final stage in the reduction of seed numbers is seen in a few rare specimens (so far unnamed), in which a large seed is attached to the base of *Glossopteris* leaves. The seed is of *Samaropsis* type, with a narrow wing — an enlarged example of the sort of seeds in *Plumsteadia*.

In South Africa, Dr Eva Endrody has described similar large seeds directly attached to the midrib on *Glossopteris* leaves. This arrangement represents the end of a reduction process from a *Jambadostrobus* fruit borne in a similar situation. (There was speculation when the single seeds attached to leaf midribs were first described, suggesting that perhaps they were simply lying on top of the leaves. However, close examination of the specimens dispels doubt about the attachment, and now the discovery of the reduction of a fruit to one seed in an Australian example confirms that this line of evolution was a feature of these plants.) 61

Petrified material of leaves and fruiting bodies of a species of *Glossopteris* discovered recently in Australia showed seeds borne on the underside of a leaf-like organ which curled around them, enclosing them in a manner analogous to the wall of an ovary, and creating a situation close to Angiospermy. The specimens were assigned to *Dictyopteridium*. However, this name should apply only to "fruits" as described and reconstructed under this name in India, in which the cover-leaf is free and the seeds are on a receptacle. Another name should be given to the Angiosperm-like fruiting body in which the seeds are enclosed by a cover-leaf. 49

144

146

147

149

148

150

149 *Nummulospermum bowense* seeds, which are the sort of seeds in *Plumsteadia ampla* and *Austroglossa* fructifications, and other Glossopterids. They are often found free in the matrix of rocks, separate from the organs which bore them. This specimen from Merewether Beach, Newcastle, New South Wales, has a group of the seeds, and their narrow, marginal wing is clearly seen. Age 255 million years. (× 1.0) 60

150 *Samaropsis* seed attached to the base of a *Glossopteris* leaf. An example of rare and interesting fossils (two others exist but, because of the preservation types, are even more difficult to photograph satisfactorily) in which the trend to reduction in the number of seeds in a fruiting body is taken to its limit, and a single large seed is seen attached to the base of a foliage leaf. This specimen from the Singleton Coal Measures at Jerry's Plains, New South Wales, age 260 million years. (× 2.2) 62

The evidence of a trend towards Angiospermy in *Glossopteris* is exciting. If combined with a trend towards reduction in the number of seeds, a situation could eventuate in which the "fruit" was a one-seeded cupule with a fused cover-leaf, and a single seeded carpel would be the result. That this would be positioned at the base of a leaf in a whorl of leaves at a stem apex is implied by the growth habit of Glossopterids. Combine this feature, then, with the already demonstrated reduction of leaves to gangamopteroid scale-leaves (analogous to petals) on the fertile tips of branches, and we have the first step to a flower. And it is only one further step to the addition of scales bearing sporangia to the same tip for the production of a bisexual flower.

GLOSSOPTERIDALES — EVOLUTIONARY TRENDS IN MEGAFRUCTI

SECTION MICROFRUCTI

Just as in the Megafructi, there are two sorts of seeds involved in the fertile structures of the Microfructi. Some scale-leaves have *Indocarpus* seeds attached, some have *Nummulospermum* type with a narrow wing all around. The scale-leaves have gangamopteroid venation.

In collections of Permian plant fossil specimens containing Glossopterids, there are a number of different scale-leaves with ill-defined venation which have seeds attached. There is no way of knowing if they are new Glossopterid forms or if they belong to unrelated plants. Other examples, such as *Partha*, are undoubtedly Glossopteridalean. 63 64 a

Two species of *Partha* are known from Australia. They are *Partha belmontensis* in which the seeds are borne in pairs on short, forking branches attached to the median line on a scale-leaf, and *Partha sessilis* in which a pair of seeds is attached directly to the margins of a scale-leaf. The seeds in both species are narrow-winged. Winged *Indocarpus*-type seeds borne on a scale-leaf in a rare specimen are a yet undescribed reproductive structure type, related to *Partha* but probably requiring a different generic name because of the fundamental difference of seed types. 65

Fertile structures consisting of a lobed head on a slender stem, known as *Rigbya arberioides*, may be Glossopterid Microfructi but the connection has never been shown. 63

No undoubted examples of *Lidgettonia*, a Microfructi type common in India and Africa, are known in Australia. This fruit type appears to be the progenitor of the forked-frond Seed-fern fructification type of the following Triassic Period. 64 d 66

151 A small scale about 1.5 cm long bearing four seeds which appear to be sessile. The venation of the scale is of Glossopterid type. From County Lincoln, New South Wales. (× 2.5)

152 Partha belmontensis, a small gangamopteroid scale-leaf, about 2 cm long, which has a pair of branches attached to it, each branch forking and bearing two large seeds. From the Insect Beds at Belmont, New South Wales. It is an example of a delicate structure which has been perfectly preserved because of the special conditions of deposition there. Age 255 million years. The genus Partha was described from India. In Australia there is a second species which has its seeds directly attached to the leaf margins, showing a tendency to reduction of seed numbers with only one pair of seeds, and simplification in attachment (Partha sessilis). (× 3.9)

153 A delicate, curving scale-leaf bears several seeds with prominent wings. Seeds are Indocarpus type. (× 3.6)

151

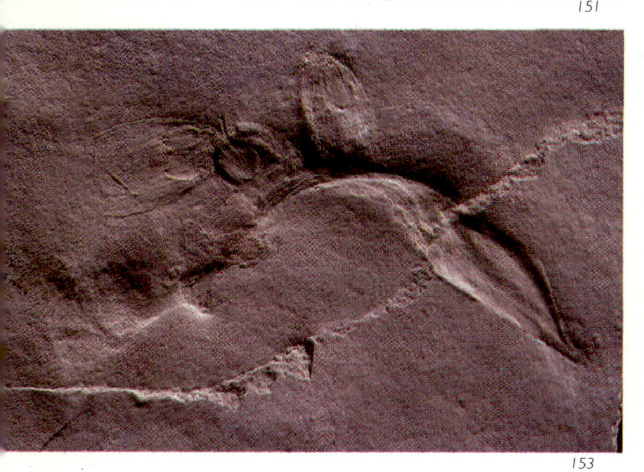

153

152

GLOSSOPTERIDALES — RECONSTRUCTIONS OF MICROFRUCTI

RIGBYA ARBERIOIDES

Possibly a fructification of Glossopterids.

154 *Rigbya arberioides.* A slender stalk, terminating in a lobed head, each lobe bearing an ovuliferous scale. Seeds are attached to scales when present, but none have been preserved in this specimen. From Newcastle, New South Wales. It has never been shown conclusively that this is a fructification of Glossopterids. No example has been found showing attachment of this sort of structure to any other fossil.

155 In this example of *Rigbya arberioides* from the Insect Beds at Warners Bay, a seed is visible, attached to one of the ovuliferous scales.

156 *"Mudgea ranunculoides"* recently described as a "tuft flower" by Dr Melville in London, is in fact an example of *Rigbya arberioides.* This specimen is in the British Museum of Natural History, and came from Ulan, New South Wales. Age 255 million years. Whether or not it is a "tuft flower" is a matter for conjecture. It could equally well be considered a cupulate structure and a forerunner of Triassic Pteridosperm fruit types. 75

MALE REPRODUCTIVE STRUCTURES OF GLOSSOPTERIDS

The only male reproductive structures proved beyond doubt to belong to Glossopterids are *Squamella* cones and scales, and *Eretmonia* and *Glossotheca* fertile scales. All have characteristic *Arberiella* striated sporangia. The mode of attachment of the sporangia to the scales varies in the different genera. In *Eretmonia* there are two forking branchlets attached to the gangamopteroid part of scale-leaves and the sporangia are on the ultimate forks, forming dense clusters. (For a long time, clusters of sporangia were only known as detached examples associated with *Glossopteris* leaves and, because of this association, were presumed to be the male phase of Glossopterids. Dr Alex du Toit in South Africa described *Eretmonia* where sporangial clusters were found attached to scale-leaves with gangamopteroid venation, further establishing the connection.)

65 67
64 c & e
67
68

157 *Eretmonia natalensis.* The scale and laminal segments of this scale-frond are clearly seen. Clusters of sporangia lie next to it. Specimen from Belmont, New South Wales, from the Late Permian Insect Beds. (× 3.0)

158 A small fertile scale of *Squamella* which had a laminar section below the scale when it was complete; with threads of sporangia attached along the base of the scale (at line of junction with absent basal part). This specimen is from the Insect Beds, Belmont, New South Wales, Late Permian age. The Insect Beds have been the source of several rare and delicate fossils because of the conditions under which the fossiliferous sediments were formed. Volcanic ash settled on the waters of a lake carrying down light, wind-blown plant debris and other floating fragments, thus enabling delicate material to survive intact. (× 1.7)

In *Glossotheca*, described from India, there are two or three pairs of forking branchlets bearing the sporangia.

Both *Eretmonia* and *Glossotheca* are still known only as detached fertile scale-leaves, but detailed knowledge of *Squamella* (whose relationship to them is shown by the same sporangia and gangamopteroid venation) implies similar arrangements for them. There is good reason to suppose that they were aggregated at the tips of fertile branches in a catkin-like arrangement (probably lax and not as closely overlapping as in *Squamella*).

Glossopteris scales (or squamae) have always been found in profusion associated with leaf fossils. They vary in shape and form: some being small, deeply concave, short and broad; others having a scale-like tip on a leaf-like lamina; and yet others being narrow and pointed. They were originally thought to be bud scales which had a protective role, forming the over-wintering buds which protect growing points during dormant seasons. They had always been found only as detached organs. Their ubiquitous association with Glossopterids and their venation with meshing veins were the reasons why they were assumed to be part of these plants.

Specimens which have been in the collections of the Australian Museum since 1892 were re-examined in 1978 and found to show the connection between squamae, leaves and stem, enabling the reconstruction of a male fertile branch of *Glossopteris linearis*.

65

As in a jigsaw puzzle, bits of evidence from a number of specimens went together to form a picture. Fertile scales were aggregated into cones or catkins at the apex of small branches. Sterile scale-leaves surround each catkin. They represent modified

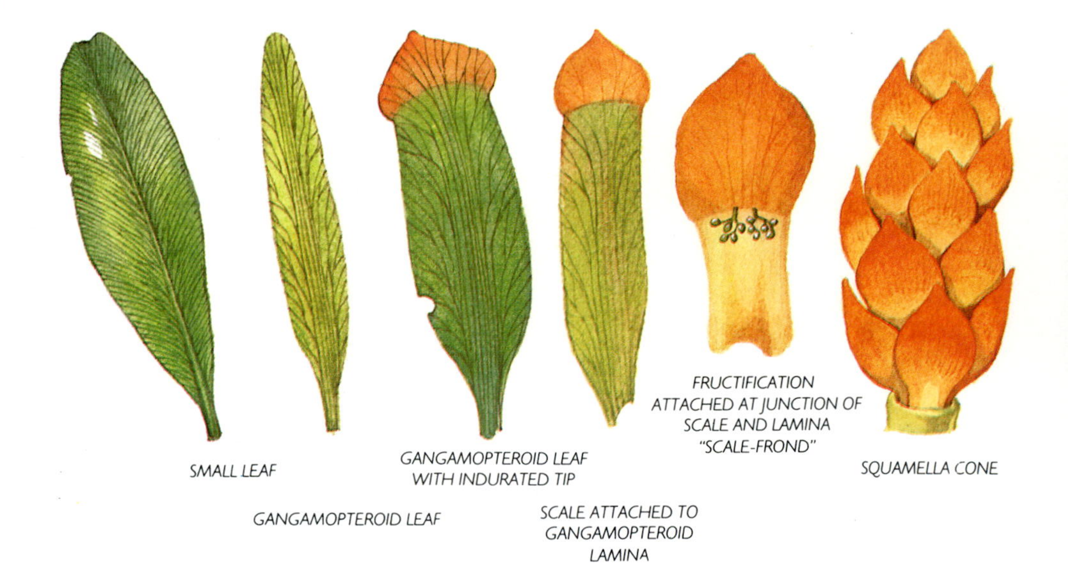

GLOSSOPTERIDALES — MALE REPRODUCTIVE STRUCTURES

159 *Squamella australis*. Enlargement of part of picture 160, photographed with a polaroid filter to show details of sporangial masses.

160 A borehole core from the Elecom Wyong bore at Tuggerah Lakes, New South Wales, was split open to reveal a mass of scales and sporangial masses from a *Squamella australis* catkin or male cone. (× 0.9)

GLOSSOPTERIS LINEARIS AND *SQUAMELLA AUSTRALIS*

161 Glossopteris linearis and Squamella australis. At the bottom of the picture is part of a whorl of leaves. One large leaf of the species, Glossopteris linearis, extends down across the specimen. The other incomplete leaves of the whorl are gangamopteroid in venation, and are modified leaves in the sequence between foliage leaves and scales bearing reproductive structures. In the centre of the leaf whorl is a scale, forming the base of a Squamella cone or catkin, which is seen broken up with scattered scales near the leaves and an aggregation of scales and sporangial masses forming a wide border at the top of the specimen. This important specimen shows the physical connection between the reproductive structures and the vegetative parts of the plant, and also shows the serial adaptation of leaves into scale-leaves and fertile scales on a stem apex, foreshadowing the evolution of floral leaves. Specimen from Flagstaff Hill, Newcastle, New South Wales. Late Permian age, 255 million years. (× 2.1)

162 Squamella sp. nov. A new species of Squamella yet to be described, from the Late Permian at Ulan, New South Wales. Age 255 million years. A stout woody axis bears spathulate scales. Large clusters of Arberiella sporangia are attached to the scales. The cone in this case is obviously a substantial structure, as evidenced by its stout axis. (× 1.9)

163 Enlargement of sporangia in picture 62. Striated Arberiella sporangia. (David Barnes)

foliage leaves and there is a series showing increasing modification from foliage leaf to smaller gangamopteroid leaf, to scale-leaves which are part indurated scale at the tip and gangamopteroid lamina below, and to fertile scales (which are mainly squamae but when found intact are seen to have a short gangamopteroid lamina attached). The catkins of overlapping fertile scales and the associated modified leaves were described as *Squamella australis*.

The fortuitous preservation of a specimen in which *Squamella* fertile scales (the base of a catkin) were surrounded by sterile scale-leaves and a number of normal foliage leaves of *Glossopteris linearis* established connection between the parts. (The rest of the catkin was preserved, broken up into its component scales on the edge of the same slab of rock.)

Another specimen showed *Glossopteris linearis* leaves attached to a stem, this time a sterile branch of the tree. Part of this stem was petrified and the wood anatomy could be studied under a microscope; *Araucarioxylon*-type wood was thus proved to bear *Glossopteris* leaves.

Evidence of the mode of attachment of the *Arberiella* sporangia to *Squamella* fertile scales came from a number of detached squamae. It was seen that unlike the arrangement in *Eretmonia* and *Glossotheca*, where sporangia are on the ultimate forks of branchlets, in *Squamella* they are attached to forking threads which are inserted on the line where the scale part of the scale-leaf joins its gangamopteroid laminal segment.

A complete *Squamella* cone of another species has recently been found in a collection from Ulan in New South Wales and is shortly to be described. Catkin-like organs composed of fertile scales have also been reported from South Africa.

69

162

163

THE GLOSSOPTERIDS AS ANCESTORS OF MAJOR PLANT GROUPS OF THE MESOZOIC, AND POSSIBLY OF THE ANGIOSPERMS

It is not difficult to see how massive Glossopterid fruits of Megafructi, comprising a receptacle with seeds attached to it (and a free cover-leaf which probably remained attached only until the fruit was ripe), can have given rise to the Cycadeoid line of evolution. Not only are the fruits like those in *Pentoxylon*, but some *Glossopteris* leaves were showing an increasing tendency to venation with prominent midrib and fine, regularly spaced, parallel lateral veins. *Pentoxylon* has occasional cross-connections between laterals, adding weight to the concept of an ancestor with meshing veins. Nor is it difficult to envisage the evolution of Bennetitalean Cycadeoid fructifications — like *Williamsonia*, with its receptacle bearing ovules and interseminal scales — from this line of Glossopterid fruits. 70

In the case of *Pentoxylon*, as well as its fruits and leaf types being previewed in the Glossopterids, the wood of the multiple steles in the trunk closely resembles that of modern *Araucaria* (and the Glossopterid wood is *Araucarioxylon*). One of the many theories of Angiosperm evolution suggests that the Pandanales (Pandanus Palm family) and other Monocotyledons were derived from the Pentoxylales, and the Dicotyledons from the Bennettitalean line, via Protocycadopsids and the Glossopterids. It is a theory not widely accepted because of the apparently general belief in the Ranales (*Magnolia*-like flowers) as being the ancestral type. It has also become generally accepted that the Angiosperms are monophyletic — that they have one ancestor. Botanists and palaeobotanists have been looking for a "missing link" to fit their theories, without success. 71 72 73

The belief in a single ancestor has to a large extent been accepted because Flowering Plants not only have their seeds enclosed in an ovary, but there is a double fertilisation process in the ovule which initiates the build-up of food reserves in the seed. This makes the seeds more efficient by providing some guarantee of survival for the next generation. It is argued that double fertilisation, which characterises all Angiosperms, is unlikely to have been developed in the same way in more than one ancestor, ruling out the possibility of polyphyletic origins. However, the argument surely loses its strength if the "ancestor" becomes an Order of plants like the Glossopterids, which have already shown so many different potential lines of development. They need only to be shown to have achieved double fertilisation to be satisfactory ancestors for several lines of evolution within the Flowering Plants. It is highly unlikely that it will ever be proved whether or not any Glossopterid had double fertilisation, so speculation will continue.

The tendency in Megafructi fruits to reduce the number of seeds and to fuse the cover-leaf and the receptacle is a possible way of forming a carpel, the basic unit of the Angiosperm ovary.

The Microfructi also show reduction in the number of seeds in their single-seeded cupules and in their sessile attachment to the scale-leaves which bore them (each unit representing an open carpel). From such fertile structures — aggregated into lax or close catkins, unisexual or united with *Squamella* male units — Amentifera, the catkin-bearing Angiosperms such as FAGACEAE (Beeches, Oaks and their allies), could have been derived.

The catkin-producing section of the Glossopterids seems to have produced a Conifer, *Walkomiella*, as one line of evolution and it seems highly likely that the Southern Conifers have evolved from Glossopterids in this category.

WALKOMIELLA AUSTRALIS 77

164 Walkomiella australis, a Late Permian conifer from Ulan, New South Wales. Age 255 million years. Branching stems of a Conifer with scale-like leaves closely clothing them in the manner of many modern Conifers. This Conifer appears in the Fossil Record towards the end of the Permian. While Conifers in general were believed by botanists to have been descendants of the Cordaites, this early one shows many features in common with the Glossopterids and may be derived from them. (× 0.8)

165,166 A specimen of a male cone and its counterpart in which carbonised material remained in the compression. Very large numbers of Arberiella sporangia were attached to the scales in this cone, and some are seen fossilised separately in the rock matrix adjacent to the cone. The venation of the cone scales is a delicate mesh as in Glossopteris. (× 2.0)

167 A female cone in this specimen with counterpart shows a line of three seeds across the middle (lower convex half). The cone consists of the expanded stem forming a funnel-shaped lower half of the oval cone structure. The sporophylls (cone scales) are inserted on the flat top of this expanded part of the stem in a tuft, and each bears a single seed at its base. (× 2.5)

168 A small branch of the Conifer, terminating in a cone. For a long time cones were only known as impressions without any fine detail. The Ulan locality is rich in examples and the very fine-grained rock in which they occur has resulted in preservation of more details. (× 4.7)

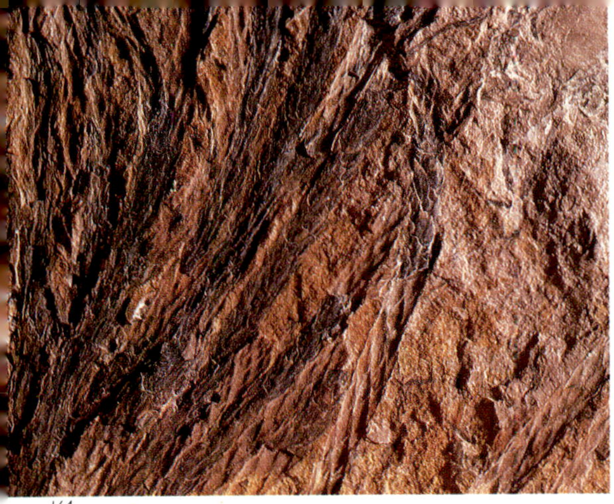

164 165

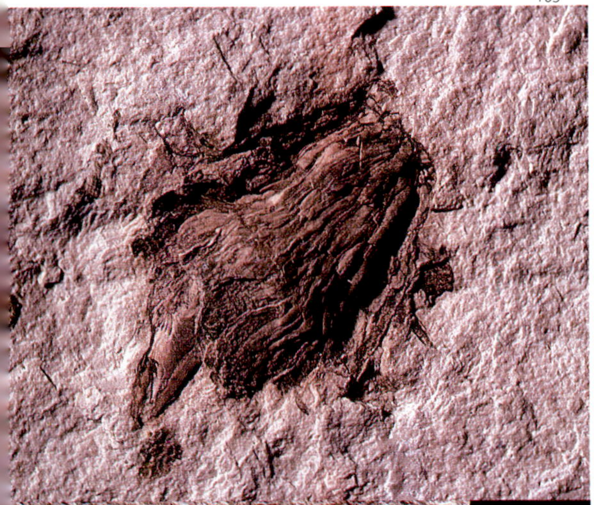

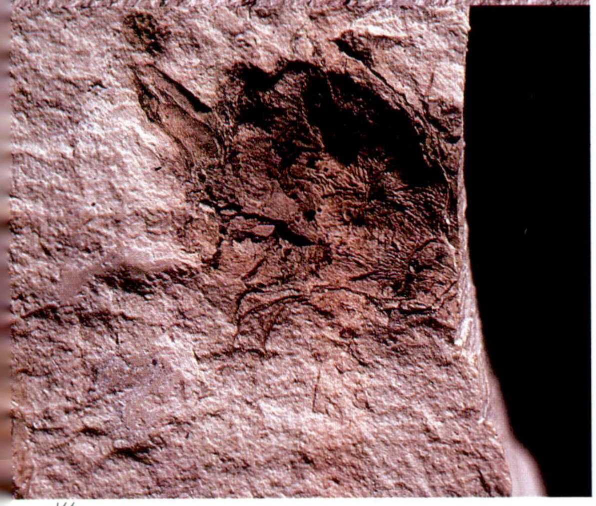

166 167 168

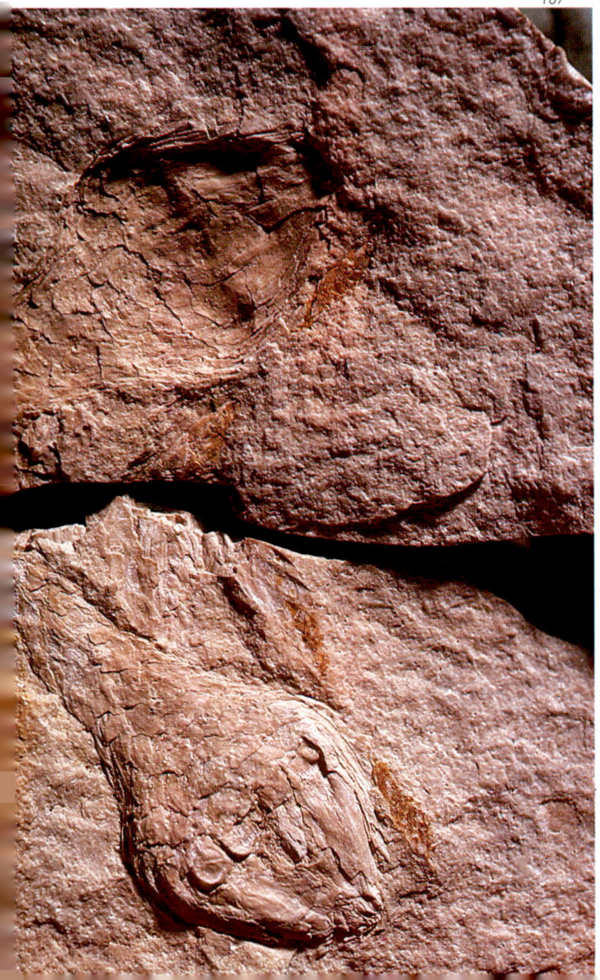

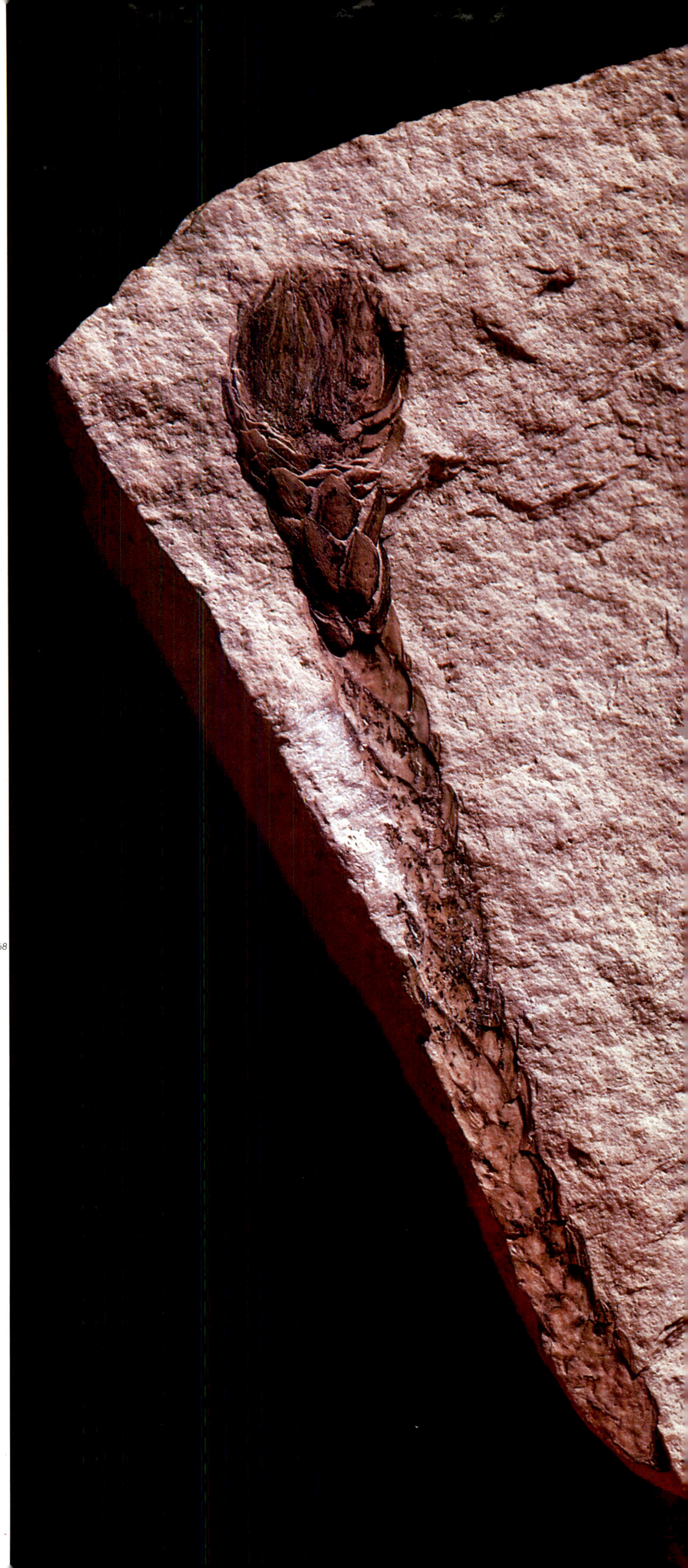

POSSIBLE LINES OF EVOLUTION FROM THE GLOSSOPTERIDALES

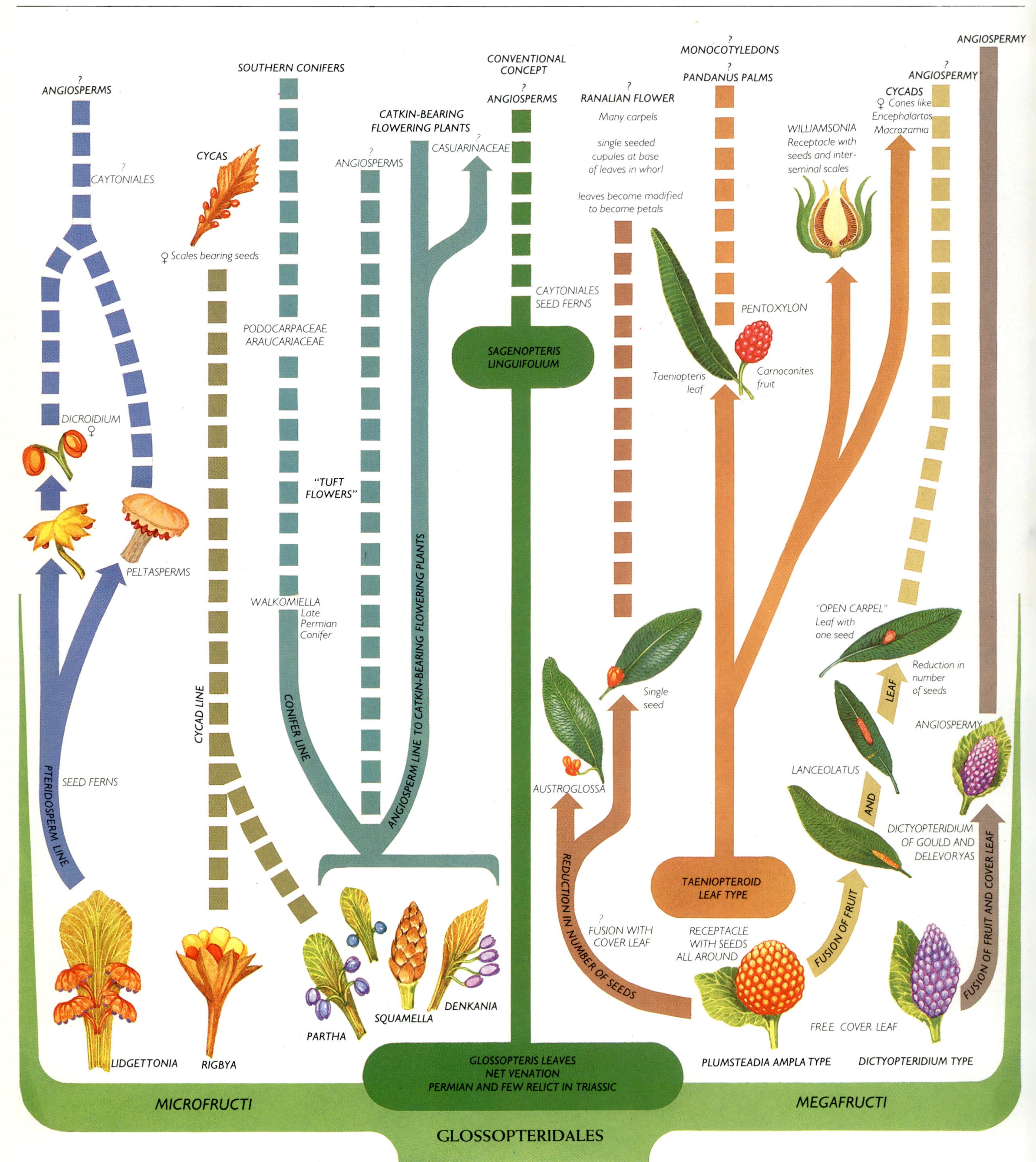

? ANGIOSPERMS
? CAYTONIALES
DICROIDIUM ♀
PELTASPERMS
SEED FERNS
PTERIDOSPERM LINE
CYCAS
♀ Scales bearing seeds
CYCAD LINE
SOUTHERN CONIFERS
PODOCARPACEAE ARAUCARIACEAE
"TUFT FLOWERS"
WALKOMIELLA Late Permian Conifer
CONIFER LINE
CATKIN-BEARING FLOWERING PLANTS
? ANGIOSPERMS
? CASUARINACEAE
ANGIOSPERM LINE TO CATKIN-BEARING FLOWERING PLANTS
CONVENTIONAL CONCEPT
? ANGIOSPERMS
CAYTONIALES SEED FERNS
SAGENOPTERIS LINGUIFOLIUM
? RANALIAN FLOWER
Many carpels
single seeded cupules at base of leaves in whorl
leaves become modified to become petals
Single seed
AUSTROGLOSSA
REDUCTION IN NUMBER OF SEEDS
? FUSION WITH COVER LEAF
? MONOCOTYLEDONS
? PANDANUS PALMS
PENTOXYLON
Taeniopteris leaf
Carnoconites fruit
TAENIOPTEROID LEAF TYPE
RECEPTACLE WITH SEEDS ALL AROUND
WILLIAMSONIA Receptacle with seeds and inter-seminal scales
CYCADS ♀ Cones like Encephalartos Macrozamia
? ANGIOSPERMY
ANGIOSPERMY
"OPEN CARPEL" Leaf with one seed
Reduction in number of seeds
LEAF
LANCEOLATUS
AND
FUSION OF FRUIT
ANGIOSPERMY
DICTYOPTERIDIUM OF GOULD AND DELEVORYAS
FUSION OF FRUIT AND COVER LEAF
FREE COVER LEAF
LIDGETTONIA
RIGBYA
PARTHA
SQUAMELLA
DENKANIA
GLOSSOPTERIS LEAVES NET VENATION PERMIAN AND FEW RELICT IN TRIASSIC
PLUMSTEADIA AMPLA TYPE
DICTYOPTERIDIUM TYPE
MICROFRUCTI
MEGAFRUCTI
GLOSSOPTERIDALES

169

170

169 Female cone of a Cycad, Macrozamia spiralis, the Burrawang, a common plant in eastern Australia, often forming an understorey in Eucalypt forests near the coast. The sporophylls are borne spirally around a central axis and bear two reflexed ovules which are protected by the tightly fitting peltate heads of the sporophylls. Cones reach considerable size, and seeds are brilliantly coloured when ripe. (Densey Clyne)

170 Pandanus Palms. Also called "screw pines" because they have leaves in three-ranked phyllotaxy inserted on twisted stems, giving a remarkable spiral arrangement. Natives of the tropics, but with a few in temperate regions, they are also remarkable for their aerial roots which form buttresses. Their flowers are aggregated into large heads, each flower results in a drupe, and the whole fruiting structure is like a large cone. They are "prehistoric" looking plants, and may be very ancient and derived from a Pentoxylon-like ancestor. (Densey Clyne)

171 A ranalian flower, the Lotus Lily, Nelumbo nucifera, grows in deep water lagoons in coastal northern Australia. There is no clear differentiation of sepals from petals. It has many free petaloid floral leaves, many free stamens and many free carpels in the centre — in this genus embedded in the surface of a receptacle, and each containing one ovule — a simple floral structure. (Buttercups and other related flowers show the same simple structure.) (Jim Frazier)

From the *Lidgettonia* strain of Microfructi, with peltate cupules bearing seeds, the Seed-ferns of the Mesozoic Era can be derived. The "flowers" of *Dicroidium* (and the reduction of these lobed forms to the helmet-shaped, single-seeded cupules of most species) and the related Peltasperms, *Lepidopteris*, are on this Pteridosperm line of evolution.

After the Permian, *Glossopteris* leaves are very rare. A few are known in Triassic strata, but the surviving leaf types which are obviously related are *Linguifolium* and *Sagenopteris*, believed to be Caytoniales, a group which has been suggested as an Angiosperm ancestor.

There has been increased interest in recent years in the possibility that Glossopterids might be ancestral to all Flowering Plants. While the foregoing account is highly speculative, it is intended to emphasise that now that palaeobotany has revealed so much about ancient plants, particularly the Glossopterids, the time has come to take a fresh look at evolution as a whole. The long accepted and often unquestioned ideas should now be questioned. There is some merit in most of the theories which have been propounded, and new concepts will arise as we become better informed about plants of the past.

74

The accepted view of evolution has been that the first Angiosperms were woody plants, probably forest trees, because the primitive Angiosperms today are those sorts of plants. However, it is being noticed, as more early Angiosperm megafossils are found, that many had very small leaves. Logic would suggest that the proto-Angiosperms may have been either small woody herbs or plants with little wood arising from root-stocks. Because they were small they would have had rapid life cycles and much better potential for evolution than trees in which each individual takes a very long time to mature. (The evolution of the Mammals from small, inconspicuous warm-blooded creatures that went unnoticed in the "forest" of large Reptiles is the same sort of arrangement.) It is also more logical to see the first Angiosperms as plants in a succession towards those which later took over the Closed Forests. By being adapted to fringe niches which were unsuited to the established ancient plants, they formed an ecological as well as an evolutionary succession towards the forest trees.

171

BLECHNOXYLON TALBRAGARENSIS 78

From Talbragar River, New South Wales. Age 255 million years. Blechnoxylon talbragarensis is a small, woody herb and because of its wood structure which is like that of a Glossopterid, the arrangement of its leaves in whorls and the bud-like structures which resemble the arrangement seen in Gangamopteris walkomii, it is not impossible that it is derived from glossopteridalean stock. When discovered the plant was described as a Fern, but the internal anatomy with gymnosperm wood and secondary growth made it highly abnormal for a Fern.

172 A whorl of leaves with a bud-like aggregation of small scales in the centre. This is a feature never seen in Ferns. (×10.6)

173 An enlargement of a microscope slide of a section of a small petrified stem of the twig in picture 175, which shows the ring of secondary wood.

174 A whorl of leaves which appear to have been fleshy with rolled margins. The venation is prominent, with strong midrib and forking laterals. (× 5.9)

175 A branching twig of this very small woody plant, whose stems are only about 1.5 mm wide and whose leaf whorls are 1 cm across. This plant has been found only once, and the specimens illustrated here are on the three small rock pieces collected in 1898. The small main stem of this specimen is petrified and microscope sections were cut and showed Araucarioxylon wood structure with secondary growth. (× 14.9)

172

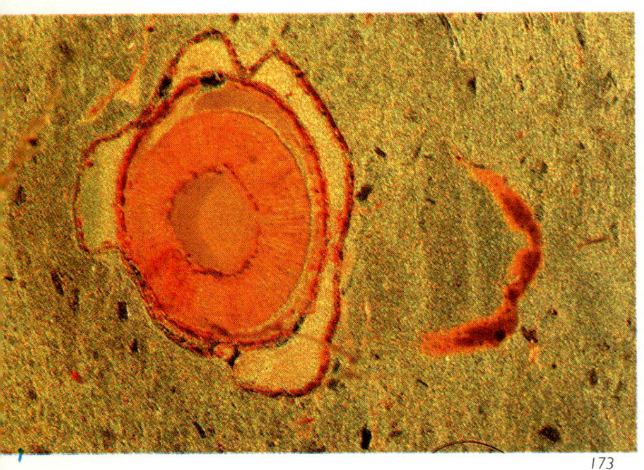

173

174

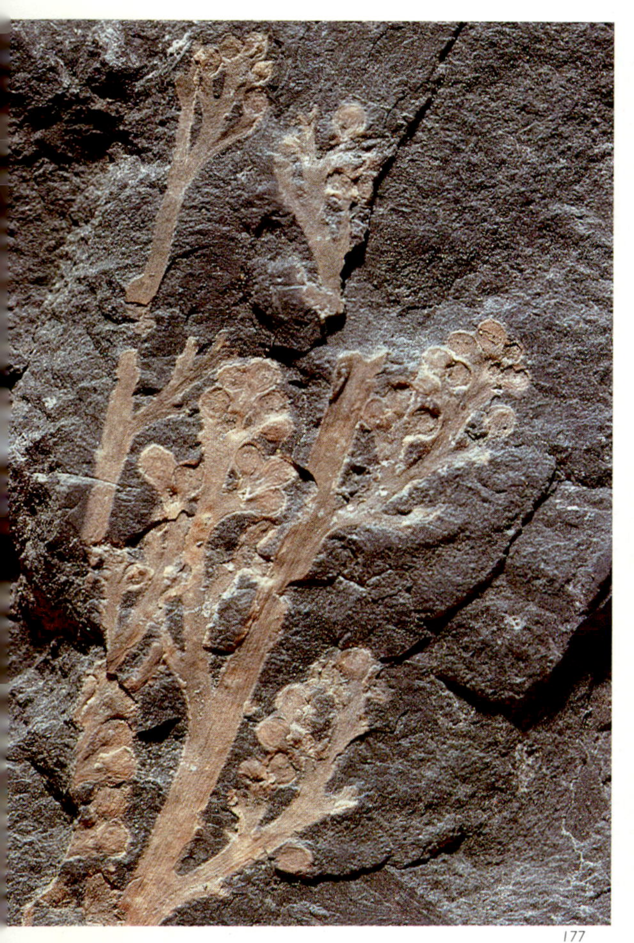

177

178

176

179

PERMIAN FERNS

176 *Sphenopteris lobifolia*, a common Fern of Permian times. From Hamilton Pit, Newcastle, New South Wales. (× 2.0)

177 Part of a fertile pinna of a Fern. From Mt Victoria Pass, New South Wales. Late Permian age, about 250 million years.

178 *Sphenopteris flexuosa*, a delicate Fern from Newcastle, New South Wales, from the Late Permian. (× 0.9)

179 *Alethopteris lindleyana* Fern foliage, probably the leaf of *Palaeosmunda* trunks. Specimen from Tryphinia, Queensland, Late Permian age. (× 1.8)

180 *Alethopteris lindleyana*, specimen from Duaringa, Queensland. (× 3.7)

PERMIAN HORSETAILS

181 Phyllotheca australis. Leaf whorls, the leaf sheath segments united near the base. These whorls surround the nodes on the segmented stems. Horsetails are known as ARTHROPHYTES, meaning jointed plants, because they are constructed of regular segments. Horsetails are common in the Fossil Record as they grew in or near water, and were thus in areas where they were likely to be included in sediments and had a good chance of being fossilised. Specimen from the Late Permian at Shepherd's Hill, Newcastle, New South Wales. Age 255 million years. (× 1.4)

182 Stellotheca robusta. A Horsetail in which the leaf sheath segments are free and elongated. From Tryphinia, Queensland. (× 2.5)

183 Umbellaphyllites ivinii. Leaf sheaths with segments united into little umbrellas. Specimen from Burning Mountain, New South Wales, which gets its name from the fact that the coal seams are on fire deep underground, and smoke emerges from cracks in the ground. The heat of the combustion has baked the shales containing the plants, and the fossils are enhanced by the dark red colour which results. (× 3.4)

184 Phyllotheca etheridgei. Leaf sheath segments again united into umbrella-like structures. From Shepherd's Hill, Newcastle, New South Wales, at which locality there is great profusion of this species. (× 2.3)

181

182

183

184

185 Noeggerathiopsis hislopi. A whorl of leaves of this Cordaite which was common in Permian times. It may have been almost as common as *Glossopteris*, but because it was not deciduous and probably grew further from water than the swamp-growing Glossopterids did, it was fossilised less frequently. Specimen from Newcastle, New South Wales. (× 0.6)

186 A small Lycopod stem showing seasonal banding, from the Late Permian at Gunnedah, New South Wales. This specimen shows that there was still marked seasonality of climate with a dormant winter season. The Permian had warmed up considerably from the ice age conditions of its inception to Late Permian times, but the climate was at best temperate with cold winters. (× 0.9)

187 A leaf of a Ginkgophyte from the Late Permian at Cooyal, New South Wales, where it is associated with *Glossopteris*. Such leaves are rare in the Permian, but as the trees probably grew on hillsides and not near the swamps, this is partly a function of the chances of being fossilised when not growing in habitats close to water. (× 1.4)

188 This pinnate frond of a Cycadophyte, *Dunedoonia reticulata*, occurs with *Glossopteris* leaves (two are seen with it at the top and bottom) in a specimen from the Late Permian at Dunedoo, New South Wales. Age about 255 million years. The reticulate venation is clearly seen. (× 2.1) 80

189 Samaropsis pincombe, a handsome winged seed, from Bar Beach, Newcastle, New South Wales. Late Permian age, 255 million years. (× 1.9)

190 Schizoneura gondwanensis. A Horsetail in which the leaf sheath segments are united into two leaves at each mode. This plant is a useful indicator of age, being confined to the Latest Permian and Basal Triassic horizon. From Newcastle, New South Wales. About 255 million years old. (× 1.4)

185

186

187

188
189
190

CHAPTER 6

THE *DICROIDIUM* FLORA

THE TRIASSIC PERIOD

FROM 245 TO 208 MILLION YEARS AGO

There was a warm and wet interval at the end of the Permian and the beginning of the Triassic. These times saw the sudden appearance of a new flora characterised by the first forked-frond Seed-ferns, Podocarp Conifers, Ferns, Ginkgos and Cycadophytes. The Glossopterids disappeared from the Fossil Record.

The climate became increasingly hot and dry as the Triassic Period proceeded, and many of the plants of the Mid and Late Triassic show drought-resistant adaptations. Dicroidium *is the characteristic plant of the Period.*

Modern style Conifers make their appearance in the Fossil Record in Triassic times.

RECONSTRUCTION OF CYLOMEIA

The climate of the Triassic was warm to hot with marked seasonality of rainfall.

It is likely that there had been a rapid changeover from the cool temperate climate of most of the Permian to a warm, moist interval on the Permian-Triassic time boundary. The climate change is evidenced by the Roof-Shale Flora (*Callipteroides* Zone Flora) at the top of the coal sequence in the South Coast coalmines in New South Wales, where there is a sudden appearance of a rich flora of Conifers and Lycopods accompanied by an abrupt disappearance of the Glossopterids.

The Early Triassic Narrabeen Group Flora of New South Wales also shows characteristics related to a warm, moist climate without marked seasonal dry periods, in the large-leaf foliose expression of the species in it. Plants in this flora do not show the clear xerophyll adaptation of the later Triassic.

The *Dicroidium* Flora of the Mid to Late Triassic shows characteristics associated with a warm to hot and dry climate.

Conditions conducive to coal formation virtually disappeared during the Triassic, except for small areas in South Australia (Leigh Creek), Tasmania, north-eastern New South Wales and south-eastern Queensland. The warm to hot climate and the marked seasonality of the rainfall discouraged plant growth in inland areas that had supported coal swamps in the Late Permian. Red-bed formation confirms the alternation of long dry seasons and rainy seasons with moderate precipitation.

In Western Australia the sea spread eastwards from Broome into an area between Geraldton and Carnarvon, before retreating in the Mid Triassic.

In the North-West Shelf area a marine transgression in the Early Triassic was followed by the establishment of a river system and the formation of deltas near the Exmouth Plateau, and by a further transgression at the end of the Triassic.

PALAEOGEOGRAPHY OF THE TRIASSIC.

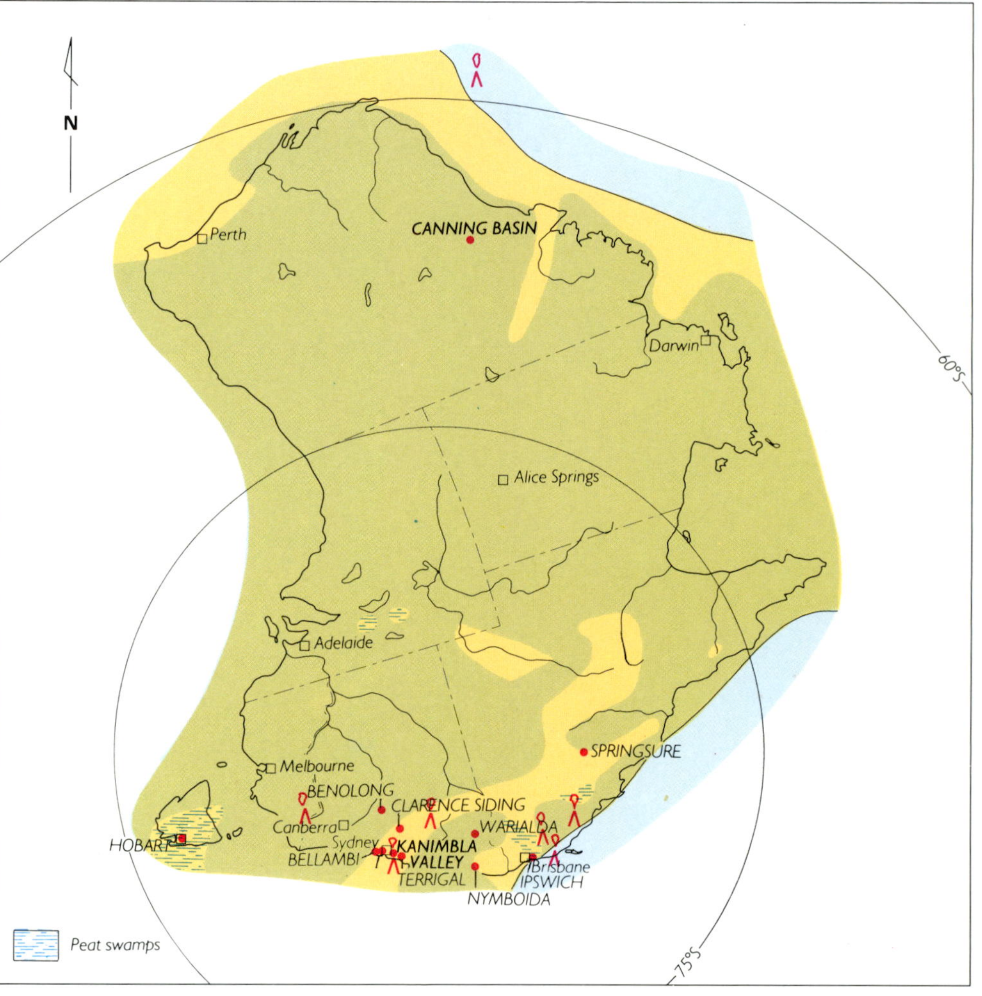

Sandstone formed during the Period constitutes some of the main reservoirs for the gas fields of the Shelf. Organic matter incorporated in the sediments, largely plant, produced the gas.

Freshwater sedimentation in the Bowen and Sydney basins of eastern Australia was a feature of the Period. Volcanoes were active on the continental shelf area and on land east of the basins.

THE ROOF-SHALE FLORA OF THE PERMIAN-BASAL TRIASSIC TRANSITION ZONE

The first of the forked-frond Seed-ferns, *Dicroidium callipteroides*, makes its appearance briefly in the Fossil Record at the stratigraphical horizon between the Permian and the Triassic. Its fronds fork repeatedly, rather than once as in the *Dicroidium* species which follow it in the Triassic. It may represent an ancestral plant which gave rise to *Dicroidium* on the one hand, and to *Lepidopteris* and the Peltasperm line of Seed-ferns on the other. Its reproductive structures are still unknown. Its ancestry is unclear.

The Seed-ferns which precede *Dicroidium callipteroides* are the Glossopterids, whose origin may have been in the aphlebiae of the Rhacopterids in the Late Carboniferous (by the suppression of the pinnate phase of their foliage). It is not too far-fetched to postulate that with a sudden change of climate a '"throw-back" to pinnate form might have occurred, and the forked-frond *Dicroidium* complex could have arisen from the "aphlebial" Glossopterids.

The Lycopods of the Roof-Shale Flora and of the Early Triassic Narrabeen Group which follows are particularly interesting. They include an Australian genus *Cylomeia*, which is different in appearance from the *Lepidodendron* model of Lycopods 81–83 characteristic of the Giant Clubmoss Flora. Instead of having bottle-brushes of leaves on the smaller branches, each plant is like a small Palm tree. The single woody trunk grew perhaps a metre high and had a crown of ribbon-like leaves flattened into a disc. The miners in the South Coast collieries refer to the fossils of the separate leaf-whorled crowns as "chrysanthemums", an inappropriate name in the botanical context but descriptive of their appearance. The root buttresses or rhizophores of the plants were lobed structures covered in the scars from stigmarian rootlets.

Reproduction was by spores from well organised cones of *Cylostrobus* type. 84 Species bearing large cones appear to have produced them at the stem apex in the centre of the leaf whorl, as in *Cylomeia capillamentum*. The smaller cones of other species probably hung out of the leafy crown on slender stems. (It is very rare to find these smaller cones with any indication of stem or attachment; the few known examples of slender stems in attachment appear to have been too thin to support the weight of the cone.) All *Cylostrobus* cones have large female spores (megaspores) on the sporophylls of the lower half of the cone and microspores on the distal sporophylls.

In the Narrabeen Group sediments of the Capertee Valley in New South Wales, small herbaceous *Cylomeia*-type Lycopods occur. Their cone bases are only about 1.5 centimetres across, but mature megaspores in the centres of the leafy discs indicate that they only attain such size.

The sudden appearance of recognisable Podocarp-type Conifers in the Roof-Shale Flora is a significant development. The modern distribution of Podocarps indicates a Gondwanan ancestry and further supports the proposition that their origin was Glossopteridalean.

STEMS AND RHIZOPHORES OF EARLY TRIASSIC LYCOPODS

191 A Cylomeia rhizophore at the base of a stem, showing its lobed form. Sometimes rhizophores are fossilised as casts in growth position in the mud in which they lived, and there is a horizon near Mona Vale, New South Wales, where such three-dimensional fossils occur. Specimen from Avalon, New South Wales. Age 235 million years. (× 1.8)

192 Two Lycopod stems with pattern of leaf base scars in vertical rows separated by ridges. Both stems have attached rhizophores, patterned with circular spots, which are attachment points for stigmarian rootlets. From Avalon, New South Wales. (× 0.5)

191

192

193 Whorls of leaves like this specimen are known as "Chrysanthemums" by miners who see them in the roof shales of the South Coast mines of New South Wales, where they are quite common. Each is the leafy crown of a small Lycopod *Cylomeia undulata.* (× 0.7)

194 *Dicroidium callipteroides*, the first of the forked-frond Seedferns, has fronds with repeated forking, unlike the characteristic *Dicroidium* of the Triassic, whose fronds fork once. This specimen is from the roof shales in the Bulli Coal Mine, New South Wales. No fertile examples of the plant are known and its relationships are not clear. The zone which it characterises, on the Permian to Triassic boundary, represents a sudden vegetation change and the virtual extinction of *Glossopteris.* A climatic change and a period of hot, wet conditions is indicated by faunal evidence as well as that of the vegetation. This period of change was from about 250 to 240 million years ago. (× 1.0)

195 *Isoetes sp.* A Quillwort type Lycopod, from roof shales at Bulli, New South Wales. Sporangia are seen attached to the base of some of the leaves (centre, lower part of fossil shows this clearly). (× 0.7)

196 Conifers *Voltziopsis africana* (left) and *Podocarpus sp.* (with large leaves) flourished during the hot, wet period of Permian to Triassic transition. The roof shales in South Coast coalmines contain a rich Conifer flora, as yet largely undescribed. Specimen from Bulli, New South Wales. Age 245 million years. (× 2.6)

194

193

195

196

197 Fronds of a Tree-fern, *Cladophlebis australis*, occur in Narrabeen sediments, some attaining considerable size. Usually only a few isolated pinnae are preserved in a specimen, but occasionally more complete fronds are found, and this photograph is of part of such a complete frond which was almost a metre long, measured along its rachis. From Turrimetta Head, near Mona Vale, New South Wales. (× 1.3)

198 *Taeniopteris wianamattae*, a Cycadophyte leaf with strong, parallel venation. From Warriewood, New South Wales. (× 1.8)

199 *Voltziopsis wolganensis*, delicate Conifer twigs with lax terminal cones and thin, needle leaves. From Kanimbla Valley, New South Wales. Age about 235 million years. (× 0.9)

200 Stems of a Horsetail, *Phyllotheca sp.*, are abundant in the Narrabeen shales of New South Wales. The plants grew as rushes along water margins and, like the Lycopods, they were fossilised *in situ* where they lived. A node is visible on the main stem, and also on the small stem below. Leaf sheaths of as many segments as there are vertical ridges in the internodes were attached to the nodes, the segments partially united at the base and free distally, or united into groups forming leaf-like units. The arrangements in leaf sheaths determine genera and species, the stems are similar in all Horsetails. From Avalon, New South Wales. (× 1.2)

197

198

199

200

CYLOMEIA, A LYCOPOD OF THE EARLY TRIASSIC

201 Part of a crown of leaves of *Cylomeia undulata*.

202 *Cylomeia capillamentum*. Two "sunflowers" comprising linear leaves in a whorl, with a large cone on the stem apex in the centre of the leaf whorl (see reconstruction). Specimen from Long Reef, New South Wales, from the Early Triassic Narrabeen Group. Age approximately 235 million years. (× 0.3)

203 Cone of *Cylostrobus sydneyensis* type, a cone of *Cylomeia*, which is a very rare example in that it has a stem attached. Cones of this sort are plentiful in the Narrabeen Group, but invariably fossilised separately, not connected to any other plant fossil. This one with a stalk shows that in contrast to the situation in *C. capillamentum*, where a sessile cone is in the centre of the leaf whorl, the thin stalk must have resulted in cones hanging down between leaves in the other species. (× 1.1)

204 Large cone, of the sort borne on *Cylomeia capillamentum*, fossilised separately in this specimen from Long Reef, New South Wales. (× 1.2)

205 A small, herbaceous species of *Cylomeia*, not yet named, with a cone base about 1.5 cm across and four Lycopod leaves (narrow, parallel-sided, with pronounced median groove) attached to the axis below the base (and part of another leaf not attached). There are mature megaspores on the segments of cone base, indicating that the plant is mature and does not attain greater size. From Capertee Valley, New South Wales, 235 million years old. (× 2.0)

201

202

204

205

203

206 Cones of *Cylostrobus sydneyensis* and long, narrow leaves. Both are part of the *Cylomeia* plant. The cones are abundant in a zone in the Narrabeen Group shales which outcrop in cliffs on Sydney's North Shore. Specimen from Long Reef, age about 235 million years old.

207 Leaf of *Cylomeia undulata*. These long, narrow leaves were originally assigned to the form-genus *Taeniopteris* which was named for its similarity to *Taenia*, the Tape-worm. They had been found only unattached until the "chrysanthemum" fossil (picture 193) was found at Bulli, New South Wales, and it was seen that they were the leaves of this small arborescent Lycopod. The undulations are probably produced by the fossilisation processes — the leaves may have been fleshy and suffered shrinkage. Specimen from Narrabeen, New South Wales. (× 1.1)

208 A Lycopod cone, *Skilliostrobus australis*, from Narrabeen Group sediments at The Skillion, Terrigal, New South Wales. The cone scales have forking projections, and the type of spores in this sort of cone differs slightly from spores of *Cylostrobus* cones. (× 0.8) 85

Reconstruction of a *Cylostrobus* cone 84

206

207

208

Reconstruction of the *Dicroidium* plant

THE *DICROIDIUM* FLORA OF THE MID TO LATE TRIASSIC

86–90

Dicroidium leaves show a bewildering range in pinnule form, with the intermediates often grading from one type into another, which creates considerable difficulty in allocating specific names. The suites of forms show increasing adaptation to arid conditions, with a decrease in the size of the lamina, thickened cuticle, and reduction of leaves to narrow spikes. Different generic names have been used for the various types, but it is now usually accepted that all are accommodated within the one genus, *Dicroidium*. Those at the end of the arid-adapted line may still be referred to *Xylopteris* but cuticle and reproductive structures indicate close relationships of all the forms, from the most foliose to the most reduced.

Female reproductive structures are of *Umkomasia* type. Again, a number of generic names have been used for the different sized and shaped types. But essentially they are all single-seeded cupules borne on the ends of the ultimate branches of a pinnate structure. They represent the modified pinnules of a fertile frond. 91 92

Male reproductive structures are *Pteruchus* type with heads or catkins of sporangia containing the spores, which have two air bladders to make them suited for long distance dispersal by wind. 93

Both male and female reproductive structures show similarity to those seen in some Glossopterids. The clusters of male sporangia are like *Eretmonia* with the scale-leaf reduced or absent, and the single-seeded cupules are like those in *Partha* and *Denkania* (without the scale-leaves) or when in "flower-like" heads are like *Lidgettonia*. 64b

209 *Dicroidium zuberi*, the commonest form of fork-frond Seed-fern. This beautiful specimen from the Beacon Hill Quarry at Brookvale, New South Wales, is about 225 million years old. (× 2.1)

210 *Dicroidium dubium* from Benolong, New South Wales, age about 225 million years, shows the characteristic forking of the genus. Its pinnae are lobed, rather than divided into pinnules. The way the frond lies on the rock and the "natural" look, like a living plant, suggests that this species appeared less stiff and fern-like than most commoner species did. (× 1.5)

210

211 *Dicroidium odontopteroides var moltenensis*, a variety of the *odontopteroides* type frond, from Benolong, New South Wales. Considerable variation in pinnule form is seen even in this one frond, some of the pinnules being spiny, others like the normal for the species, and some intermediate. (× 1.5)

212 *Dicroidium sp.*, a delicate, leafy frond, different from most common forms of the genus in its appearance. There is a wide variety of forms, and inter-grading between species, and it is often difficult to classify individual specimens. (× 1.3)

213 *"Xylopteris" tripinnata*, a drought-adapted form of *Dicroidium* in which the pinnules are reduced to spiny protrusions. From Beacon Hill Quarry, Brookvale, New South Wales. The Triassic was hot and periodically arid, and many plants show xermorphic adaptation. Thick cuticles, which are often preserved as a result of their substantial nature, are common in *Dicroidium*. (× 3.4)

214 *Dicroidium odontopteroides* (right), a very common type of leaf throughout the Triassic, with pinnules more rounded and less wedge-shaped than *D. zuberi*, is associated in this specimen with *Dicroidium spinifolium*, a reduced form with leaves like elongated spines. From Ipswich in Queensland, where the Denmark Hill locality and others in Ipswich Series shales are famous for their beautiful fossils. Age Late Triassic, about 215 million years. (× 2.4)

211

212

213

215 *Dicroidium odontopteroides* and *Linguifolium diemenense* (the wedge-shaped leaf, centre left) from Newtown, Hobart, Tasmania. Note the similarity of this *Dicroidium* specimen to specimens from Antarctica and South Africa illustrated on pages 38 and 39. (× 3.2)

216 *Dicroidium odontopteroides*, a handsome, forking frond, from Ipswich, Queensland. From the Late Triassic, about 215 million years old. (× 1.1)

217 *Dicroidium odontopteroides* with the frond of a Cycadophyte, *Pterophyllum nathorsti*, across it. This beautiful specimen is from near Springsure in Queensland, in the Great Artesian Basin. The pinky-mauve sediment and the grey impressions are an unusual and fortunate combination. (× 1.8)

218 *Dicroidium cuticle*. The spots on the surface are the bases of glandular hairs which are scattered all over the leaf surface.

215

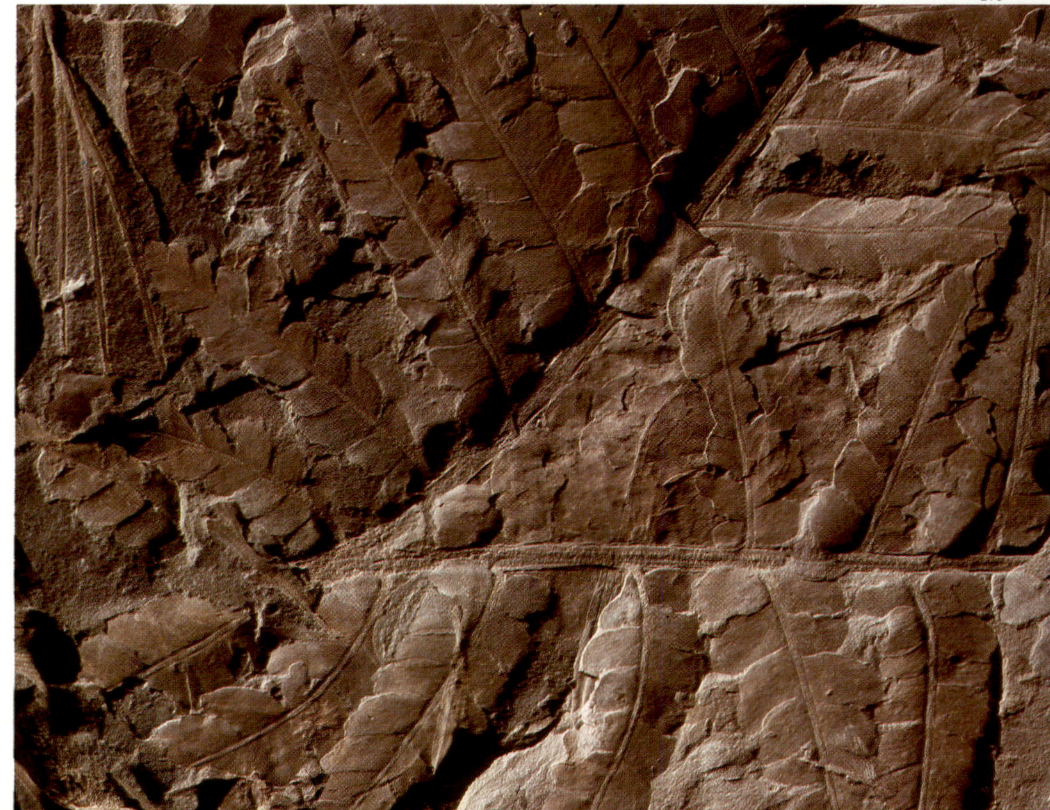

216

INSECTS ASSOCIATED WITH *DICROIDIUM* FLORA OF SOUTH AFRICA DURING THE TRIASSIC

86b

A 1980s review of the Late Triassic Molteno Formation in South Africa lists Insects which occur with plant fossils at a number of localities. As similar assemblages of Insects would have been present in all parts of Gondwana, this information is of interest in the Australian context.

	Family	Genera	Species	Individuals
Ephemeroptera (Mayflies)	1	1	1	5
Odonata (Dragonflies)	1	1	3	16
Blattodea (Cockroaches)	2	3	5	110
Orthoptera (Grasshoppers)	3	3	3	11
Homoptera (Bugs)	6	7	7	16
Mecoptera (Scorpion Flies)	1	1	1	11
Lepidoptera (Butterflies)	1	1	2	3
Coleoptera (Beetles)	3	4	6	61

These Insect fossils are mostly of isolated wings. The scaly-winged *Lepidoptera* are of particular interest as they are ancestral to the Butterflies, which are usually only associated with true flowers and whose further evolution is related to flower structure evolution.

217

218

219 A female fertile branch of a *Dicroidium* which has seed-bearing structure instead of leaves. From Lorne Basin, northern New South Wales. (× 1.5)

220 *Pteruchus barrealensis,* a male catkin of *Dicroidium,* which contains the microsporangia and male spores. Specimen from the Lorne Basin, northern New South Wales. Age 225 million years. (× 1.2)

221 *Umkomasia,* female reproductive bodies of *Dicroidium.* These helmet-shaped structures are single-seeded cupules. Specimen from Benolong, New South Wales, Middle Triassic age, about 225 million years.

219

220

221

222

222 *Dicroidium* "flowers" from Clarence Siding, New South Wales. Before it was known how *Dicroidium* reproduced, this specimen had been described as *"Williamsonia"* flowers" and was considered to be a Cycadophyte reproductive structure. It was also described as leaf sheaths of a Horsetail. (This information is given to highlight the problems facing the palaeobotanist who has to rely on how things look to classify them, often without many facts to substantiate determinations.) (× 4.2) 92

223 "Flowers" of *Dicroidium*. These palmate structures bear the seeds under the lobes, and are the female reproductive organs. Specimen from Mt Piddington, New South Wales. Age Triassic. (× 2.7)

223

OTHER PLANTS OF THE *DICROIDIUM* FLORA

Dicroidium is associated with other Seed-ferns (like *Lepidopteris*), Tree-ferns, Ginkgophytes, Cycadophytes, Conifers and Horsetails.

LEPIDOPTERIS

94 a b

Fronds of *Lepidopteris* look superficially much like *Dicroidium*. Their female reproductive structures are different. The name of the family to which they belong is PELTASPERMAE, which refers to the peltate (or shield-like) structures which carry the seeds. Radially symmetrical peltate heads, with a central stalk and seeds on the undersides of the segments of the head, characterise this group of Seed-ferns. This type of structure recalls the *Lidgettonia* type of female structure in the Glossopterids.

Male structures are known as *Antevsia* and are similar to *Pteruchus*.

FERNS

Ferns were abundant, growing in suitable sheltered habitats as they do today. Many were Tree-ferns attaining considerable size, as evidenced by petrifactions of trunks. Delicate ground Ferns were also present.

GINKGOPHYTES

Ginkgos become common in Triassic floras. They show a range of forms, from those in which the leaf blade is dissected into a few broad segments to those in which the segments are linear and arranged like a fan. Others have an undivided strap-like form.

Great abundance of Ginkgos in some horizons indicates that they were locally dominant in some woodland habitats.

CYCADOPHYTES

Cycad-type leaves start to appear in the Fossil Record during this Period. The reproductive structures of the plants which bore these leaves are not known, but the cuticular structure of the leaves indicates a relationship with Cycads. Some are assigned to the form-genus *Taeniopteris* (or *Macrotaeniopteris*, in the case of very large leaves) and have long, narrow, parallel-sided form and lateral veins are parallel, running straight to the margins at right angles to the prominent midrib. The name *Taeniopteris* is given because of the long, thin shape like *Taenia*, the Tape-worm. An interesting Triassic Cycadophyte of another kind is *Cycadopteris scolopendrina*, whose colourful and descriptive specific name means "Centipede".

CONIFERS

95 96

Conifers of several general types are common in the Triassic Period. Podocarps of the genus *Rissikia* are particularly abundant at some localities. Others with small scale-type leaves like the *Walkomiella* of the Permian and the needle-leaved *Voltziopsis* occur less frequently. Because the Conifers lived on the dry hillsides, somewhat remote from water, they are less likely to have been included in any macrofossil sample and are always under-represented.

HORSETAILS

Horsetails were common throughout the Triassic Period. They are particularly abundant in the Narrabeen Group sediments, which were laid down in extensive deltas where the Horsetails were growing and could be fossilised *in situ*.

Some leaf forms of Ginkgophytes.

224 A Lycopod rhizophore from the Moolayember Formation at Springsure in Queensland, showing that Lycopods were present in the flora. Of Triassic age. (× 1.2)

225 Lycopod stems with widely spaced leaf base scars, from the Erskine Sandstone in the north-east Canning Basin of Western Australia. (× 0.7)

226 Ginkgoites semirotunda, a Ginkgophyte leaf with its blade divided into ten free segments, free to the base. From Benolong, New South Wales, where there is a beautifully preserved Dicroidium flora, with abundant Ginkgophytes. Middle Triassic age, about 225 million years. (× 5.0) 89

227 Cycadopteris scolopendrina, a robust Cycadophyte frond which presumably reminded its author of a Centipede. A rare plant which has only been found in a brick pit at St Peters, Sydney, in Ashfield Shale, and which was collected in about 1880 and has not been found anywhere else since. About 225 million years old. (× 2.6)

224

225

226

227

HORSETAILS OF THE TRIASSIC

228 *Neocalamites hoerensis*, a stem with long, very narrow leaves (segments of leaf sheath) attached to the nodes. Specimen from the Beacon Hill Quarry at Brookvale, New South Wales, which was worked out and filled in, and thus no longer exists. A very fine collection of animal and plant fossils was made from the quarry over the years. (× 2.9)

229 Horsetail stem with an attached cone. Preservation is poor as the sandstone is grainy, and detail is not preserved, but it can be seen that the cone has a central axis with sporangiophores and was attached at a node of the stem. The specimen is from the Triassic Blina Shale in the northeast Canning Basin of Western Australia. (× 0.7)

230 Horsetail stems with very prominent branch scars at nodes. From the Erskine Sandstone, in the north-eastern Canning Basin, Western Australia. (× 1.4)

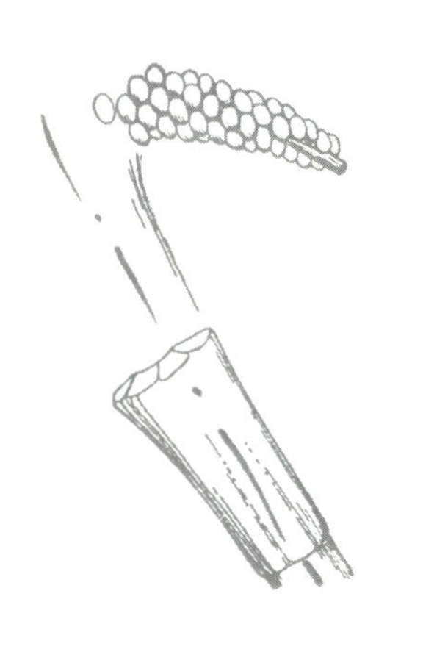

228

230

229

LEPIDOPTERIS STORMBERGENSIS

231 A large frond of this Fern-like plant from Benolong, New South Wales. From the Mid Triassic, about 225 million years old. Unlike a Fern, this frond has pinnules attached to the rachis between the pinnae.

232 The fruiting body of Lepidopteris is a shield-like organ, hence the name Peltasperms for this group of plants. This specimen shows the lobed head of a fertile structure. Seeds were borne underneath each lobe, and the pedicel was central. From Benolong, New South Wales.

Reconstruction of a peltate structure

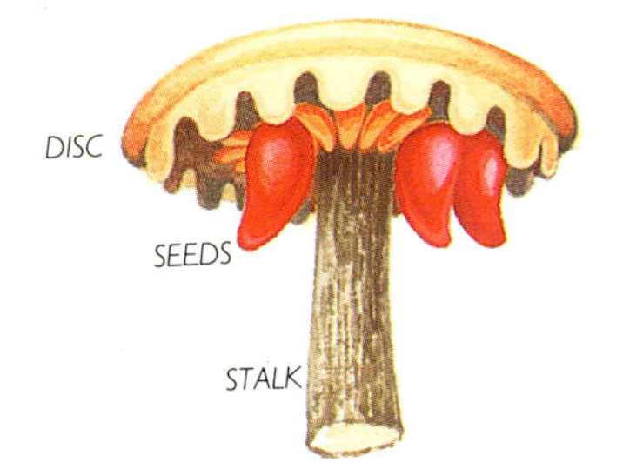

231

232

233 A fertile frond, probably of a Fern, from the Beacon Hill Quarry at Brookvale, New South Wales.

234 Rienitsia spatulata, portion of a large Fern frond from Calga Quarry, Gosford, New South Wales. (× 1.1)

235 Todites pattinsoniorum, part of a Tree-fern frond. From Benolong, New South Wales. Middle Triassic age, about 225 million years. Tree-ferns were abundant during Triassic times. (× 1.8) 89

236 Rissikia media, a Conifer, and Phoenicopsis elongatus, a Ginkgophyte, from Nymboida, New South Wales. (× 1.5)

CHAPTER 7

THE AGE OF CONIFERS AND CYCADS

A modern Kauri Pine, Agathis, *very similar to its ancestors of 175 million years ago.*

THE JURASSIC PERIOD

FROM 208 TO 144 MILLION YEARS AGO

The Jurassic Period was uniformly warm to hot and wet worldwide, and there was a luxuriant cosmopolitan flora of Conifers, Cycads, Ferns, Seed-ferns, Ginkgos, and herbaceous Lycopods and Horsetails. The flora, which continues into the Early Cretaceous, is the last to be composed of plants from ancient groups only. After the Jurassic, the changeover to modern-aspect flora commenced.

Worldwide, the climate of the Jurassic Period was uniformly warm and wet. There were no polar ice caps and there was apparently little differentiation into climatic belts. World floras had similar composition. It was the age of ancient forests of Conifers, Cycads and Ferns, equivalent to the warm temperate Rainforest.

Australia still lay in high latitudes, between 60° and 45°S, but even adjacent Antarctica carried forests of Conifers and Cycads at this time.

In the Early Jurassic, major river systems developed in the east of the Australian continent, flowing generally eastwards and entering the sea at a point north of Brisbane. These rivers spread a sheet-like sand body across the area. This sandsheet, after burial and minor folding, is now one of the main aquifers of the Great Artesian Basin (the Hutton Sandstone). In places it forms a reservoir for the accumulation of oil and gas.

Basalts were extruded near the western margin of the Sydney Basin. In Western Australia coal swamps existed north of Perth and there was a variety of marginal marine and freshwater environments on the North-West Shelf.

In Mid to Late Jurassic times alluvial deposition on the eastern part of the continent continued as the river systems expanded. Some rivers may have flowed north through the area of the present Gulf of Carpentaria. Others drained part of Western Australia, flowing into the sea between Port Hedland and Broome.

There were coal swamps in north-eastern New South Wales and south-eastern Queensland, possibly along the lower reaches of an east-flowing river.

A major geological event occurred when parts of Gondwana flanking the Western Australian coast started to drift away, commencing from the north-west. There had been faulting and rifting over some millions of years preceding this.

PALAEOGEOGRAPHY OF THE EARLY JURASSIC.

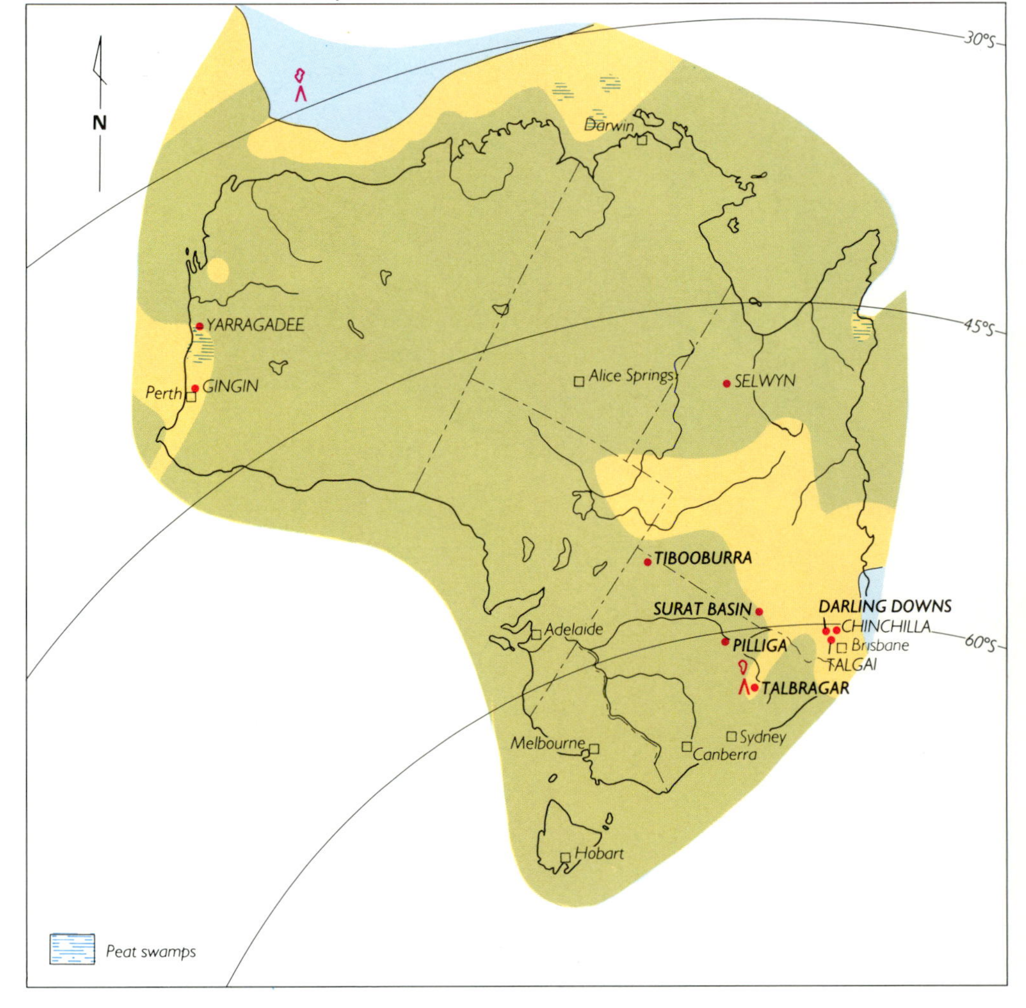

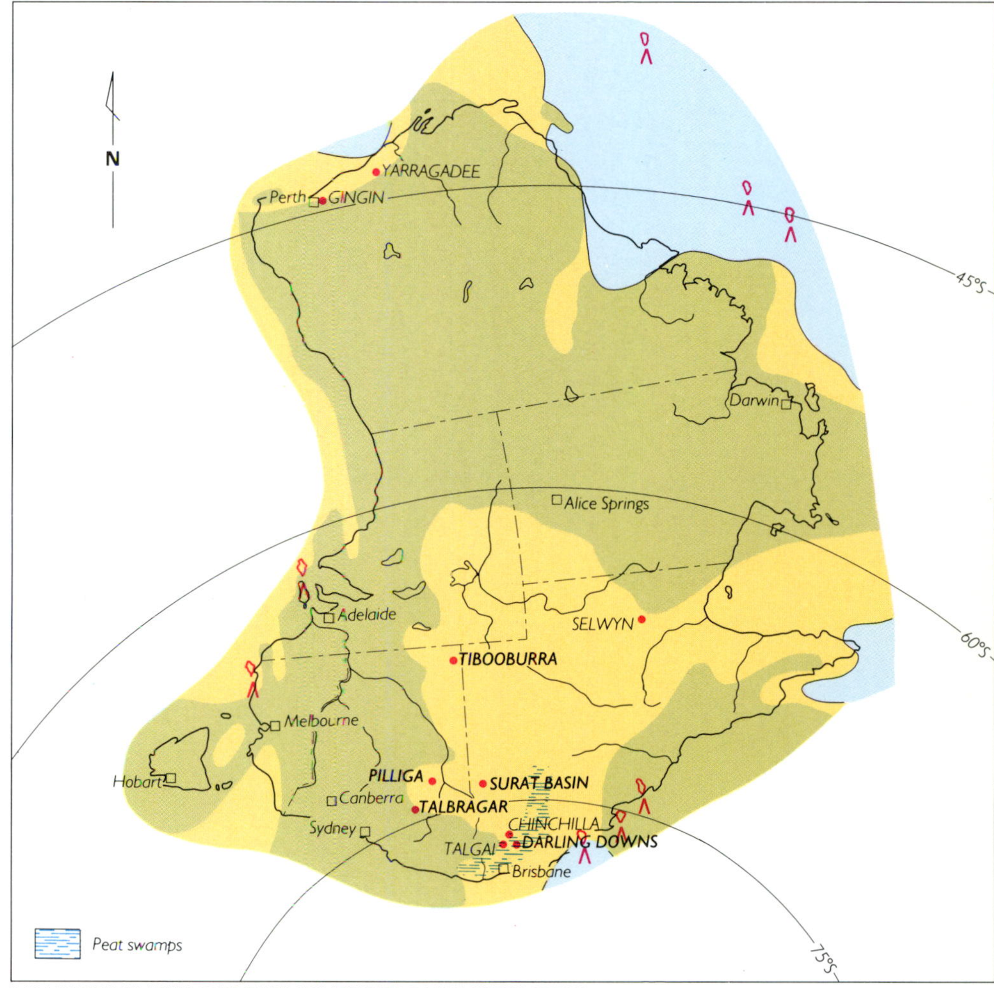

PALAEOGEOGRAPHY OF THE MID JURASSIC TO EARLY CRETACEOUS.

In the Mid Jurassic, similar faulting and lake and river deposition started on Australia's southern margin, accompanied by igneous activity. The extrusion of basalts on Kangaroo Island and west of Melbourne was associated with this first phase of Gondwana break-up. Basalts probably covered most of Tasmania at this time, but only occasional dolerites have survived the subsequent erosion.

THE TALBRAGAR FISH BED FLORA

97 98

Australia's most famous Jurassic flora is that which occurs in the Talbragar Fish Beds near Gulgong in New South Wales. Associated with the very numerous Fish, and excellently preserved as white impressions on the ochre-coloured silicified shale, is an interesting assortment of plants. Because of the beauty of the specimens they have been much sought after by collectors as well as by scientists.

The Talbragar Fish Beds deposit is of limited extent and outcrop. It represents an accumulation of fine-grained sediments in a lake which existed for a relatively short time about 175 million years ago. Its still waters, and the anaerobic conditions in the fine iron-rich silt on its bottom, provided ideal conditions for the preservation of the Fish and plant remains which accumulated there. When the deposit was discovered in 1889 the Geological Survey of New South Wales arranged for a professional museum collector named Charles Cullen to make an enormous collection from the site. We are in his debt to this day, as this comprehensive collection is so large and so complete that the subsequent ravages of collectors at the site have not meant the loss to science that would otherwise have been the case. Considering the remoteness of the location in those early days, and the fact that

237 A Fish, *Leptolepis talbragarensis*, from the Talbragar Fish Beds.

237

238 A beautiful cone, Rissikia talbragarensis, which is the cone of a Podocarp. Attached to the pedicel (right of cone base) are two long, narrow leaves, which serve to identify the type of foliage from the different varieties of Conifer leaves found in the Fish Beds. An Agathis twig, unrelated to the cone, appears in the top right hand corner, and a Fish, Leptolepis talbragarensis, is in the lower left corner of the picture. In the cone the seeds are reflexed, with micropyles towards the axis. (× 2.2)

transport of tonnes of rock had to be accomplished by bullock wagon, the collecting was no mean feat, and Cullen camped at the site for weeks.

Museums have acquired considerable collections from the Talbragar Fish Beds since that first one was gathered, but much valuable material is in private hands and unavailable for study. By the 1970s the deposit was almost worked out and at this late stage it was declared a Geological Monument, with the intention of preserving what remains.

From the plants, we know that the Talbragar lake lay in a forest of Kauri Pine (*Agathis*) interspersed with Podocarp Conifers. In the understorey was the Cycadophyte *Pentoxylon australica*. The genus *Pentoxylon* appears only briefly in the Fossil Record and is a possible ancestor for the *Pandanus* Palm line of modern monocotyledonous Angiosperms. The heath zone round the lake was occupied by descendants of the forked-frond Seed-ferns (*Dicroidium*). Tree-ferns and ground Ferns grew in suitable habitats as they do today.

This reconstruction of a landscape for the Talbragar Fish Beds lake of 175 million years ago is particularly interesting because relict *Agathis* forests on the Atherton Tableland have Podocarps growing among the Kauri Pines and a tree-like Cycad, *Lepidozamia hopei*, is present in the understorey. Thus, we see a modern assemblage of plants with the same basic composition (though at a more evolved stage) and we know exactly what the Jurassic vegetation looked like.

AGATHIS JURASSICA

239 This 175 million years old Kauri Pine (Agathis jurassica) shows its affinities with modern Agathis in the form of its leafy shoot. Leaves are long and narrow, and bud scales remain at the base of the shoot representing a year's growth. Kauris are today confined to relict Rainforests in Australia, but were widespread in the Jurassic when the climate was uniformly warm and wet. (× 1.3)

239

240 Foliage of the cone *Rissikia talbragarensis* (238). Long, narrow,

(× 2.7)

241 Cone scale of *Agathis jurassica*, an Araucarian Conifer. (× 1.9)

242 An immature cone of *Agathis jurassica*. Modern Kauri Pines shed many immature female cones, which lie in the litter of twigs and mature cone scales beneath the trees. (× 2.2)

243 Foliage of the *Elatocladus planus* form-species. When the full affinities of a fossil are not known because there is no evidence of reproductive organs to enable a determination to be made, the specimen is called by a name which describes it and which may later be found to include unrelated plants which merely have leaves which look similar. Ferns and Conifers often have similar leaves in quite unrelated genera. (× 1.6)

240

241

242

243

244 A cone of a Seed-fern, possibly of *Pachypteris*. The small umbrella-like sporangiophores are spirally arranged on the axis. (× 2.1)

245 A small Podocarp Conifer, with stems clothed in scale-like leaves, and three cones at the tip of each branch. (× 1.3)

246 A specimen which shows most of the components of the Fish Bed Flora on one slab. (× 0.3)

247 *Pachypteris crassa*, a Seed-fern, and a descendant of the fork-frond Seed-ferns of the Triassic. (× 1.7)

244

245

247

246

PENTOXYLON AUSTRALICA

Reconstruction of *Pentoxylon*

248 A long, narrow leaf of *Pentoxylon australica* at left of specimen, associated with an *Agathis jurassica* twig. Leaves of this sort were known as *Taeniopteris spatulata* before they were known to be part of the *Pentoxylon* plant. (The *Agathis Jurassica* fossils were known as *Podozamites lanceolatus* and originally considered to be Cycadophytes) (× 1.9)

249 The female fruit of *Pentoxylon australica*. The photograph is of a rubber mould made by pouring latex into a deep impression with the outline of a cone in a specimen, and thus obtaining a cast of the object which made the impression. The hard, woody seeds, with a keeled ridge, originally had fleshy covers, and the whole fruit was rather like a mulberry. The name *Carnoconites* was given to unattached fruits of this sort in India before they were known to be part of the *Pentoxylon* plant. (× 2.5)

250 Male sporangiophore of *Pentoxylon* (See reconstruction of the whole plant to understand the relationships of different parts.) Male branches bore a corona of sporophylls which were thin axes bearing a number of large sporangia containing the micrōspores. (× 4.3)

249

250

251 Petrified *Pentoxylon* wood. Section of a trunk of the plant, which shows the multiple wood cylinders. Because there are often five, the name *Pentoxylon* (from *pent*, meaning five, and *oxylon*, meaning wood) was given. Each cylinder shows annual rings in its wood. This specimen is from Chinchilla in Queensland as the type of preservation at Talbragar does not result in petrification, but only in impressions. (× 0.8).

251

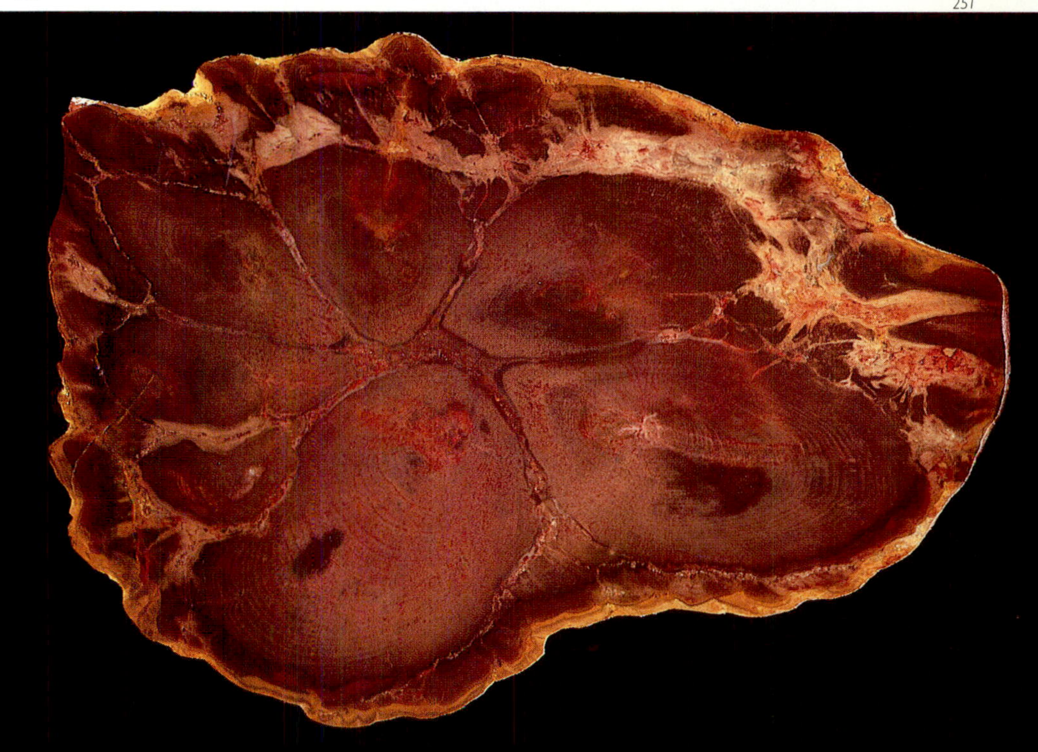

OTHER JURASSIC PLANTS

Elsewhere in Australia there are many localities where plant fossils of Jurassic age occur.

Petrified woods are an important component of these floras. Tree-ferns are particularly abundant in some localities and at places such as Tibooburra in New South Wales pieces of petrified Tree-fern trunk may be found scattered on eroded land surfaces. Many of these specimens are partly opalised and very beautiful when cut and polished.

In the Chinchilla district of Queensland, woods of *Pentoxylon* showing the characteristic arrangement of several vascular cylinders with annual rings inside each trunk are commonly found. Petrified Conifer wood is common, both Araucarian (like modern Norfolk Pine wood) and Podocarp type.

Conifer foliage of several different sorts is present in all Jurassic sediments and occasionally cones or cone-scales are found.

Cycadophyte fronds of several kinds — *Otozamites*, *Pterophyllum*, *Ptilophyllum* and *Zamites* — are also often found.

Ferns are common and of many different types, some ground Ferns and some Tree-ferns.

The Lycopods of this Period are all small plants. There are *Selaginella*-like delicate herbaceous ones, and much more robust *Isoetes*. The latter occurs as a cast of the stem and rhizophore of a whole plant, and as leaves and megasporangia, from Gingin in Western Australia. Living *Isoetes* are the Quillworts, and Tasmania has representatives of this genus growing in swamps.

Sagenopteris leaves with net-venation similar to leaves of Glossopterids require special mention. They belong to a group of Seed-ferns called Caytoniales, which may have evolved from the Glossopterids and have been suggested as possible Angiosperm progenitors.

252 *A microscope slide of a thin section of a Tree-fern trunk, showing the traces to the fronds.* (× 5.2)

253 *Pentoxylon australica. Transverse section of compound trunk which has six major wood cylinders. From Chinchilla, Queensland.* (× 1.6)

254 *An opalised Tree-fern trunk, cut and polished to show the beautiful pattern of leaf traces — the vascular supply to leaves. Specimen from Tibooburra, New South Wales, where blocks of petrified Tree-fern can be found lying exposed on eroded land surfaces near Mt Stewart.* (× 1.4)

255 *Pentoxylon australica. Oblique section through a compound trunk resulting in a "tiger stripe" pattern in the wood. From Chinchilla, Queensland.*

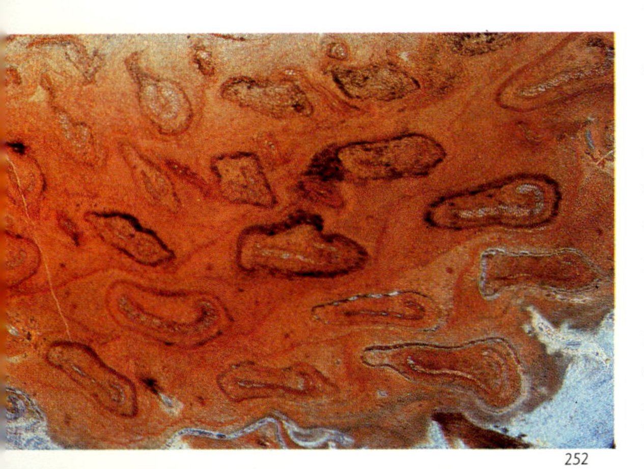

252

253

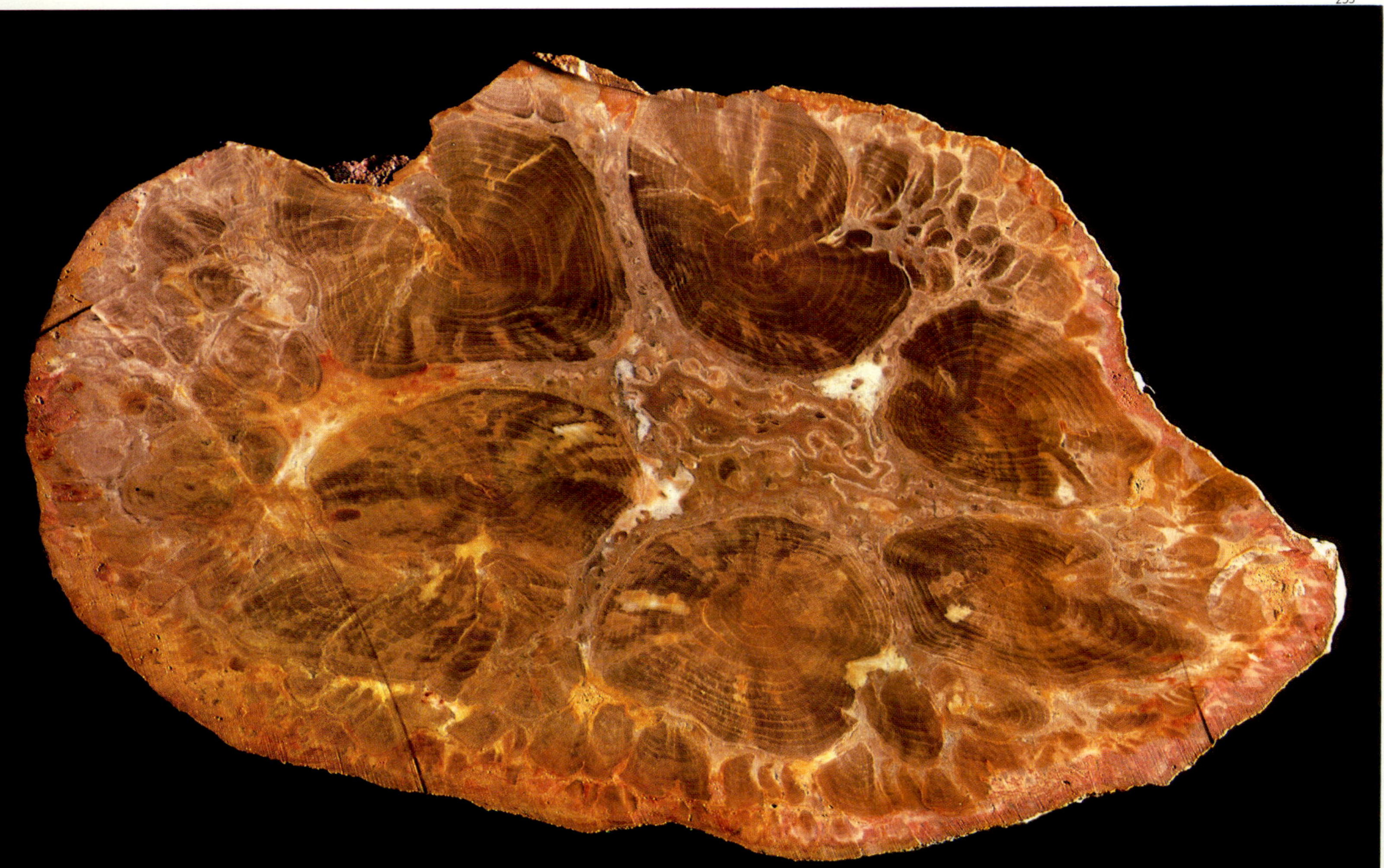

256

FERTILE JURASSIC FERN

256 Aspleniopteris sp., a rare example of a fertile Fern. The sterile foliage is seen at the top middle and the right in the picture, and the fertile frond is at the left. This is a very small and delicate Fern, seen much enlarged in the illustration, and the fertile rachis at the extreme left is only 2 cm long. The pinnae are reduced and large synangia (compound structures containing many sporangia) replace them in the fertile state. Preservation is so good in this very fine grained rock that the individual components of some of the synangia are visible. From Selwyn, in northern Queensland. Age approximately 175 million years.

NORTHERN TERRITORY CYCADS

This "prehistoric plant", *Cycas media*, is a direct descendant of Cycads of the Jurassic and Early Cretaceous, when these plants and other Gymnosperms dominated world floras.

Relict communities of Cycads and Palms, such as occur at Palm Valley in Central Australia, are a source of wonder and surprise. They live in limited environments where the microclimate still favours them, having changed little from the environment of the distant past — oasis-like refuges in an arid land.

In suitable niches in the well-watered tropical north, Cycads are an integral part of the vegetation. Burrawangs are an under-storey in Eucalypt forests in southern coastal New South Wales, and they also colonise sand close to the sea. *Lepidozamia hopei* in relict Kauri Pine (*Agathis*) forest fills the niches that were occupied by *Pentoxylon* in Jurassic Kauri Pine forests 175 million years ago.

257

257 Cycas media.

258 *Phlebopteris alethopteroides*, a Fern frond from the Darling Downs in Queensland. (× 1.3)

258

259

260

ISOETES, A FOSSIL QUILLWORT

259 Stem and rhizophore of a Quillwort fossilised as a cast. The rhizophore (bottom of cast) shows a pattern of spots where stigmarian rootlets were attached. The stem is squat and wide with a patterning from fallen leaf bases. From Pilliga, New South Wales. (× 0.9)

260 Megasporangia of *Isoetes*. Large megaspores can be seen in these sporangia, which are about 0.5 cm long and are here shown much enlarged. From Gingin in Western Australia. (× 5.5)

261 *"Taeniopteris spatulata"*, a very common leaf in Jurassic strata, which is now known to be the leaf of *Pentoxylon australica* from evidence in the Talbragar Fish Bed Flora. Specimen from the Surat Basin, Queensland. (× 3.2)

262 *Otozamites bechei*, Cycadophyte fronds from Yarragadee in Western Australia. (× 0.7)

263 *Sagenopteris rhoifolia*, Glossopterid-like leaves of gymnosperms believed to have descended from Glossopterids on the Seed-fern line. They have well marked midribs, arched laterals with cross-connections, and are classed in Caytoniales, a plant order which is believed to be ancestral to Angiosperms. From Talgai, Queensland, about 175 million years old. (× 1.3)

262

261

263

JURASSIC CONIFERS

264 An Araucarian Conifer from Yarragadee in Western Australia. The similarity of this fossil to Hoop Pine foliage *(Araucaria)*, a living representative, is obvious. The Araucarians and Podocarps are Southern Conifers and may have had a different origin from other Conifer groups, being derived from the Glossopterids. (× 3.8)

265 *Bellarinea barkleyi*, a Podocarp from Bex Hill, New South Wales. (× 1.5)

266 A delicate scale-leaf Conifer from the Roma district, Queensland. (× 2.0)

264

265

266

CHAPTER 8

DAWN OF THE ANGIOSPERMS

THE CRETACEOUS PERIOD

FROM 144 TO 66.4 MILLION YEARS AGO

Following the long Jurassic Period with its uniformly warm and wet climate and its thriving vegetation dominated by Conifers, Cycads and Ferns which had continued into earliest Cretaceous times, there was a sudden change in conditions and flora. Rapid cooling and the start of the Gondwana break-up caused a eustatic sea level rise, inundating vast areas of land and heralding the new world of continents on the move and the birth of modern-style plants which were to come to dominance in due course. The ancient plant groups, which had comprised the floras up to that time, were senescent and did not have the genetic potential to cope with the rapidly changing environment. As the climate warmed up again and the epicontinental seas retreated in the Early Cretaceous, the stage was set for change and a steady progression towards the modern world.

When the first Angiosperms evolved in West Gondwana, in the rift valley between the hump of Africa and northern South America, contact with the Euramerican landmass enabled the first radiating waves of Flowering Plants to spread into the Northern Hemisphere. An unbroken Gondwana allowed them to spread out to all the southern continents. As continents moved apart and links were severed, descendants of the first Angiosperms could spread only where migration routes remained open. The sequence of events in the Gondwana split-up determined the distribution of the ancient families of Flowering Plants. Distinct modern floras of the various continents are products of evolution in isolation since separation of landmasses — the northern into America and Europe-Asia, and the southern into its separate lands.

PREVIOUS PAGE
Pandanus Palms in silhouette against the dawn. (Jim Frazier)

After the uniformly warm and wet conditions of most of the Jurassic, there was a sudden general cooling, and probably polar glaciation. The eustatic sea level rise of the Early Cretaceous may have been in part due to melting of the ice cap, as well as to the sea floor spreading which accompanied the start of the Gondwana break-up. There is good faunal evidence in Australia that the sea adjacent to the south-eastern coast was cool to cold, and that there was general warming of ocean waters by the Late Cretaceous. Ice-rafted boulders and dropstones are found in sediments which accumulated in the inland sea during the global rise in sea level. They are evidence of either glaciers on high mountains or winter icing of rivers and shallow waters.

The changed climatic conditions of the Cretaceous Period are of vital significance in the evolution of the vegetation. The ancient forests of Conifers, Cycads and Ferns were under stress. Flooding of vast tracts of land on all the landmasses of the world resulted in denudation of parts of the landscape. As the seas retreated they left new, bare land to be colonised. Thus there were the ingredients for change: pressure created by climatic factors, and space ready to be moved into as plants suited to the new niches evolved. The ancient groups of plants that had dominated the scene up to that time did not have the flexibility to meet the challenges. But the Flowering Plants did. Their progenitors had been waiting in the wings and they had the genetic potential to explode onto the scene now that the time was right.

A similar situation existed in the animal world. The reign of the Reptiles was coming to an end and the Mammals were coming into their own. Mammal progenitors had also been inconspicuous members of communities since Triassic times and were now ready to conquer the land. Birds were likewise poised, ready to take to the skies.

PALAEOGEOGRAPHY OF THE EARLY CRETACEOUS.

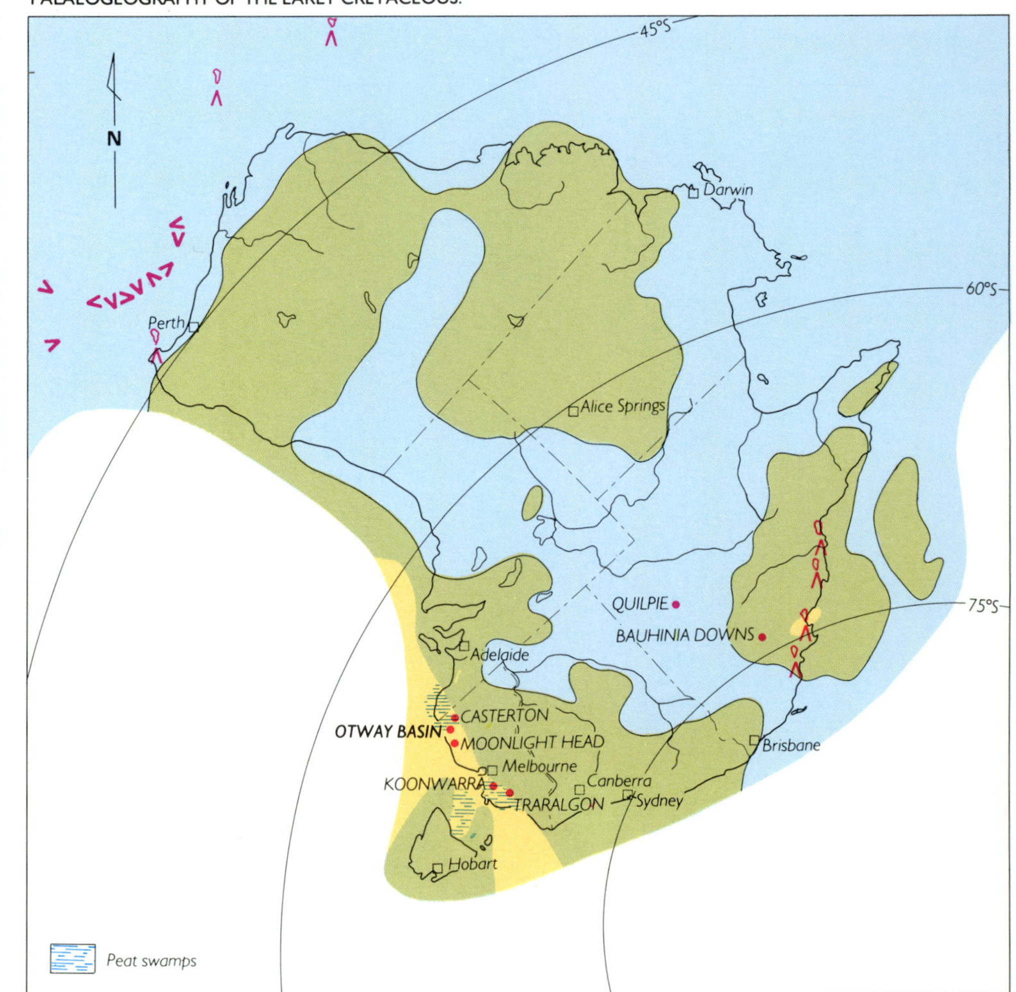

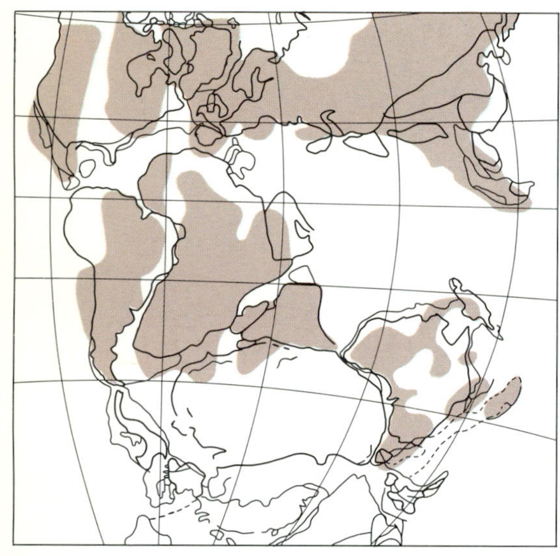

GLOBAL SITUATION IN EARLY CRETACEOUS.
Probable areas of dry land.

GREAT ARTESIAN BASIN.
Extent of Cretaceous Sediments.

The global rise in sea level in the Early Cretaceous flooded the interior of Australia, separating it into four island blocks. The fact that so much of the continent was flooded shows that it was already well planated. The Eromanga Basin of western Queensland, adjacent New South Wales, the Northern Territory and South Australia became the Eromanga Sea, and an arm of the sea extended into the eastern part of Western Australia. Flooding of the western part of the Gulf of Carpentaria resulted in local concentrations of manganese in the sediments, forming the rich Gove deposits. In the north Eromanga and south Carpentaria basins the transgressing seas were very shallow and supported a voluminous growth of floating Algae. The oil shale deposits of the region owe their origins to the organic matter from these algal communities. On the western side of the Australian continent, sea floor spreading was occurring near Carnarvon and Perth as the Gondwana break-up gathered momentum. On the southern margin of the continent, thick sediments began to accumulate in basins that were formed by rifting which preceded separation from Antarctica. In Bass Strait these sediments included coal.

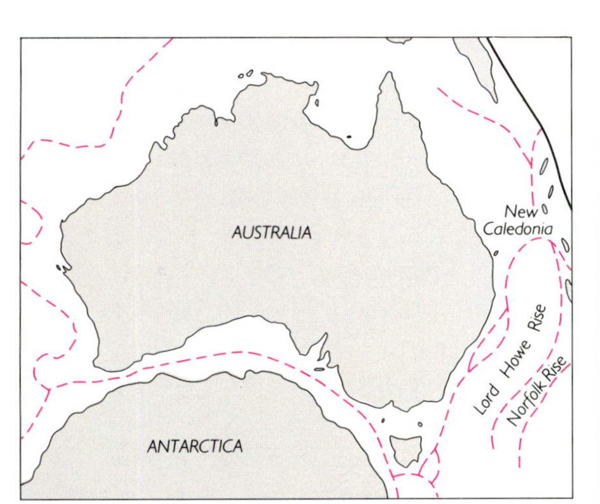

PRE-DRIFT POSITIONS OF NEIGHBOURING FRAGMENTS OF GONDWANA IN LATE CRETACEOUS AND EARLY TERTIARY.

PALAEOGEOGRAPHY OF THE LATE CRETACEOUS.

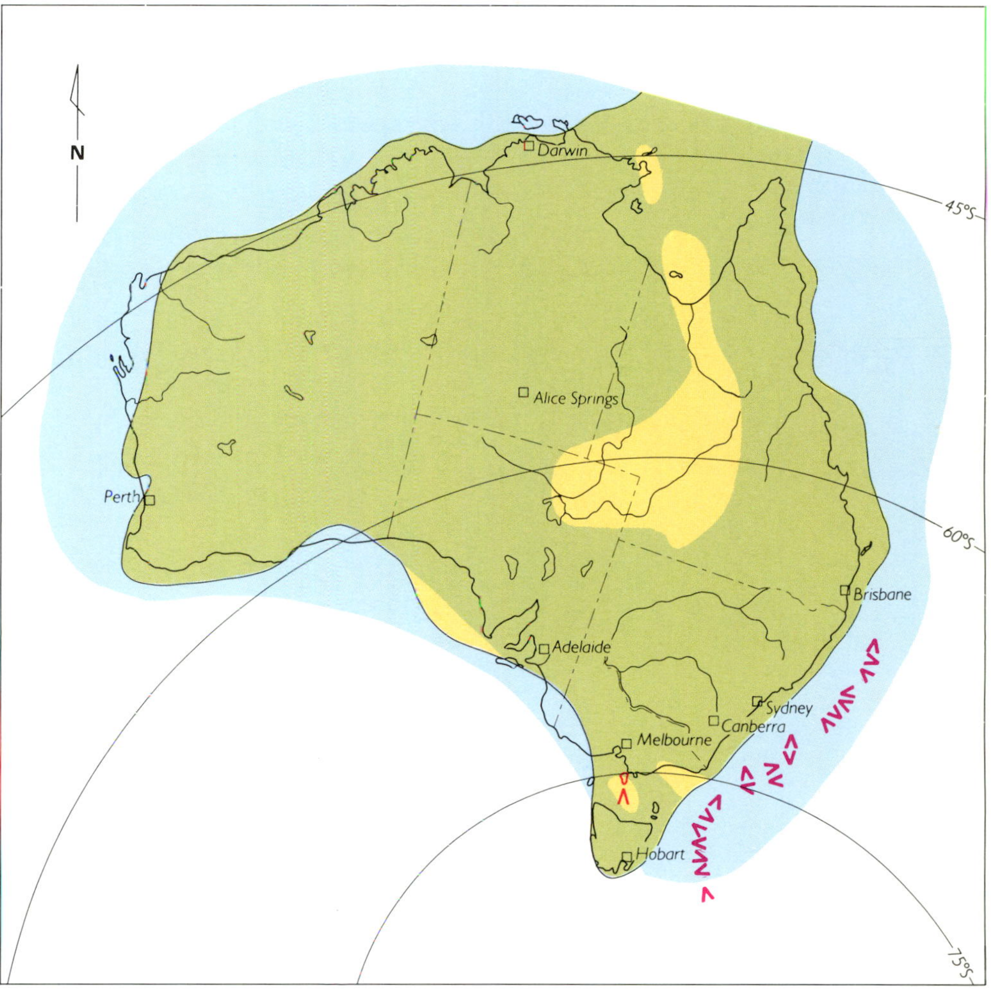

Towards the Late Cretaceous the seas had retreated from most of the continent and the climate warmed up. Sediments which had accumulated in the Eromanga Sea became the sandstone sequences of the Great Artesian Basin. Tectonic events associated with the opening of the Tasman and Coral seas resulted in uplift of the eastern edge of Australia, forming the steep seaward slope of the Great Divide. A great depth of continental sediments continued to accumulate in the basins between Tasmania and Victoria. Those in the Gippsland offshore area contained plant remains which were to be converted into the major oil and gas reserves of that

ABUNDANT INSECTS IN THE CRETACEOUS

The Koonwarra Fish Beds of Victoria have a rich Insect fauna comprising mainly aquatic Nymphs and some Water Beetles.

The Beetles were particularly numerous and probably paralleled the modern situation by having a great deal of interaction with Cycads.

Burrawang Cycads, for example, have many species of Weevils always present in their cones. The Weevils lay their eggs in the soft scales of immature female cones and eat the spores of male cones. Their significance as pollinators has been realised only recently as their activities are nocturnal and had not been observed. They may be accidental pollinators simply by reason of their coming and going in and out of the cones in which they live and feed, but their activities are of obvious advantage to their hosts. They may also have value to the plants, which grow in poor sandy soils, through their production of nitrogenous waste products as well as organic matter on their death.

267

267 A Veliid Water Bug. (Peter Jell)

KOONWARRA FISH BED PLANTS

268 Rienitsia variabilis (Bottom of picture), a Seed-fern, with leaves of a Podocarp, Rissikia sp. (top of picture). Age approximately 125 million years. (× 4.9)

region. The terrestrial origin of these fossil fuels is of particular interest as comparable development of hydrocarbons elsewhere in the world is of marine origin.

By the Late Cretaceous a new fracture line had developed along the south-east coast, and the Lord Howe Rise, which is now mostly submerged, began to drift eastwards, initiating the formation of the Tasman Sea. Formation of the Tasman Sea moved New Zealand and New Caledonia to their present positions relative to Australia, and the distances separating these landmasses have remained constant since the Early Cainozoic.

As in the Jurassic, Conifers, Cycads and Ferns characterise the vegetation. The macrofossil record is incomplete, with only Early Cretaceous floras represented. Although rare Angiosperm-like leaves have been recorded in some assemblages, their true affinities are unknown, and it is only in the Pollen Record of Late Cretaceous strata that there is real evidence of the arrival of the Flowering Plants.

Victoria has a famous Early Cretaceous flora whose composition indicates cool, wet conditions. Abundant Ferns, Mosses and Conifers with Ginkgophytes and Cycadophytes are present. There is evidence that the Koonwarra Fish Beds, in which the flora is well represented, owes its rich Fish fauna to winter-kill conditions due to freezing over of lakes and rivers in the cold winters. 100

A beautifully preserved Early Cretaceous flora from the Northern Territory comprises plants which indicate warmer and drier conditions. Faunal evidence indicates that there was climatic zonation and that the southern parts of Australia were colder than the north-western regions. 101 102

In the Northern Territory flora are examples of *Williamsonia* "flowers". These female cones of Cycadophytes have a bulbous receptacle, or head, to which its seeds were attached, and the whole structure was surrounded by bracts. The "flowers" were produced by Palm-like plants similar to *Pentoxylon* in growth habit, with leaves of *Otozamites*, *Pterophyllum* and *Ptilophyllum* form-genera types. There is evidence that such Cycadophyte "flowers" had inter-relationships with Insects. Some of them have been found to have nectaries (or honey glands) on their bracts, presumably for attracting Insects in the same way as nectaries function in Flowering Plants.

On the world scene, the Flowering Plants first appear in the Fossil Record in Early Cretaceous sediments, about 140 million years old, in West Gondwana. These fossils are proto-Angiosperm pollen grains. Such pollen cannot be assigned to any living families or genera, but is of Magnolioid, Lauralian and Lilioid types. (None of these very early types occurs in Australia.) 103 104 105

From these simple types rapid evolution and diversification took place, presumably on the fringes of the contracting Conifer forests which were under stress with the changes in the climate and the increasing patterns of seasonality of rainfall with dry periods. The ancient families of Angiosperms which evolved in West Gondwana spread out into both Hemispheres.

The largest concentrations of living primitive Angiosperms on Earth today are found in Australia's northern Rainforests. Their presence in such numbers has led to the assumption by some botanists that the Flowering Plants originated in the Australian part of Gondwana. However, because the Pollen Record shows otherwise, the concentration of primitive forms in Australia is now known to represent a relict flora.

The Gondwana split-up sequence during the Cretaceous and the Early Tertiary is of great importance because of its influence on the migration and dispersion of the Angiosperms.

269 Sphenopteris sp., a delicate Fern frond. From the Koonwarra Fish Beds. (× 7.0)

NORTHERN TERRITORY EARLY CRETACEOUS FLORA

270 Pterophyllum fissum, a Cycadophyte. The fronds in this species are somewhat irregularly divided into segments (or pinnules) and segments often have a double-apex tip, hence the species name, "fissum", which means divided. From Bauhinia Downs, Northern Territory. Age about 125 million years.

271 Extremely fine twigs of a scale-leaf Conifer, each about 1 mm wide. These very fine branches of an unidentified Conifer are very common on all the specimens collected at Bauhinia Downs, and yet there are no larger Conifer branches present. (Compare with modern Cypress twiglets.) (× 3.2)

272 A Fern, Microphyllopteris gleichenioides, which closely resembles modern Gleichenia, a common Fern with a widely known species growing in damp areas on Hawkesbury Sandstone in the Sydney region, frequently seen in road cuttings. (× 1.5)

273 Otozamites bechei, a Bennettitalean Cycadophyte frond. (× 2.3)

270

271

272

273

274 *Araucarites sp.*, a twig of an Araucarian Conifer. Koonwarra Fish Beds, Victoria. (× 1.9)

275 Forking pinnae of a Fern, *Amanda floribunda*, and a pinnule of *Phyllopteroides dentata*. (× 1.8)

276 *Taeniopteris daintreei*, a Cycadophyte leaf with pronounced midrib and lateral veins at right angles to it and parallel to each other, closely crowded. From the Koonwarra Fish Beds.

277 A petal-like structure, affinities unknown, from the Koonwarra Fish Beds. (× 2.3)

278 An example of *Phyllopteroides lanceolata*, from Quilpie in Queensland. Early Cretaceous age, 125 million years. (× 3.2)

279 *Ginkgoites australis*, from the Koonwarra Fish Beds, is a Ginkgophyte with the leaf deeply dissected into a number of segments. (× 3.9)

280 *Phyllopteroides dentata*, a leaf type which for a long time was thought to be a Pteridosperm, but is now believed to be a pinnule of a compound leaf of an Osmundaceous Tree-fern. From Devil's Kitchen, Victoria. Age about 125 million years. (× 3.6)

274

275

276

277

278

279
280

SEQUENCE OF EVENTS IN THE GONDWANAN BREAK-UP AND ITS EFFECT ON THE FLORA

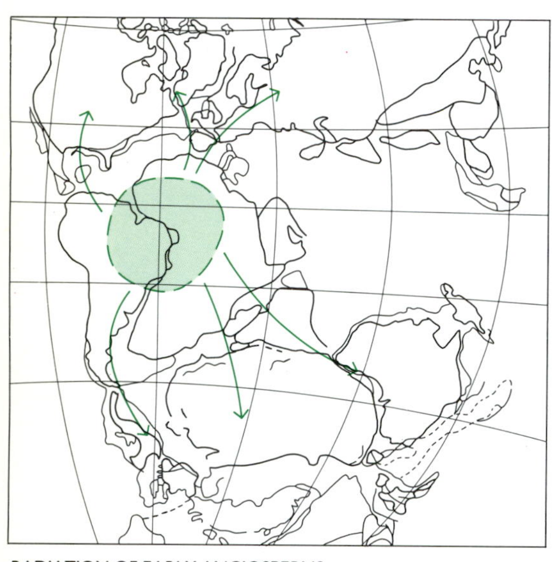

RADIATION OF EARLY ANGIOSPERMS.
Early Cretaceous reassembly showing West Gondwana as centre from which first Flowering Plants could spread North and South.

At the time when the first Angiosperms were evolving and dispersing, Turkey, Iran and Tibet had already moved north. Sea-floor spreading was starting to separate Africa from Madagascar-India. Africa, with Madagascar and India still effectively attached, was involved in rotational separation from South America and Antarctica. Africa and South America were sliding past each other along a complicated series of transform faults and the southern tip of Africa was being transformed past the Falkland Plateau. A migration route from Africa, across India to Australia probably remained open until the Mid Cretaceous.

106

The South Atlantic Ocean was formed as Africa and South America moved apart, and increased in width as the movement progressed. At first the seaway was closed to the north. Migration from Africa to Brazil was probably possible until the Mid Cretaceous, which would account for similarities in the floras of Africa and South America.

Africa-Madagascar-India became separated from South America-Antarctica-Australia-New Zealand as the rotational movement proceeded. About 90 million years ago India started moving rapidly northwards. It collided with Asia in the Mid Eocene, about 50 million years ago, and the impact pushed up the Himalayas.

The route from South America, through Antarctica to Australia and New Zealand remained open for the Palaeoaustral Flora of *Nothofagus* and its associated plants.

WILLIAMSONIA, A CYCADOPHYTE "FLOWER"

281 Hollow cast of a young Williamsonia "flower". (× 0.8)

282 Cast of a complete Williamsonia "flower". A large spherical receptacle is surrounded by elongated bracts. The base of the cast shows the opening into the hollow receptacle, which bears seeds and scales on its outer surface. Some organs of this sort had nectaries to attract Insects. From Bauhinia Downs, Northern Territory. Age 125 million years. (× 1.4)

281

282

GONDWANA BREAK-UP SEQUENCE — 1.
135 million years ago.

PALMS IN AUSTRALIA

Palms were among the earliest Angiosperms. They became established in a "Palmae Province" in global Equatorial regions with the first radiations of Flowering Plants. Some ancestral Palms reached Australia while a route existed across India (up to about 100 million years ago) and from them our endemic Australian Palms have evolved in isolation.

Lawyer Vines, wickedly armed with curved thorns, are Gondwanan examples which are little changed from the early ancestral Palmae.

There is also a northern element of Palms which have invaded the tropical parts of Australia since contact was made with South East Asia.

283

283 *Sand Palms in the Northern Territory.*

Africa is not part of the *Nothofagus* province. The genus presumably arose in temperate parts of Gondwana after Africa had become effectively isolated.

In Africa, palynological evidence suggests climatic zonation into two provinces after the initial dispersal of the Angiosperms in the Mid Cretaceous. The northern province was warm and arid and was part of the "Palmae Province" which was widely distributed in low latitudes in the Northern Hemisphere by the Late Cretaceous. The southern province was characterised by the Podocarp (Southern Conifer) vegetation type, with Angiosperms becoming dominant by the end of the Cretaceous.

The Pollen Record for the Late Cretaceous and the Early Tertiary in South Africa shows a close relationship to the Australian flora, due to the same Gondwanan ancestry. The same families dominated the vegetation. Pollen of PROTEACEAE is abundant. *Casuarina* pollen appears from the Palaeocene to the Miocene, although the genus is absent from Africa today, and is confined to Australia. RESTIONACEAE, RUTACEAE, THYMELAEACEAE, MYRTACEAE, EUPHORBIACEAE and many other Angiosperm families which are part of the Australian flora are present. The Southern Conifers, ARAUCARIACEAE and PODOCARPACEAE, are abundant as in Australia at that time.

107

Until the opening of the Tasman Sea, which occurred between 80 and 60 million years ago, New Zealand and the Lord Howe Rise formed the outer edge of Gondwana. By the time separation had been accomplished and the distance between Antarctica, Australia and New Zealand was great enough to create a barrier to plant distribution, the major diversification and radiation of the Flowering Plants had occurred. New Zealand took with it a full complement of the Palaeoaustral Flora.

This flora, the Antarctic Beech-Southern Conifer assemblage which survives in circum-Antarctic regions today (in South America, the Falkland Islands, Kerguelen,

GONDWANA BREAK-UP SEQUENCE — 2.
95 million years ago.

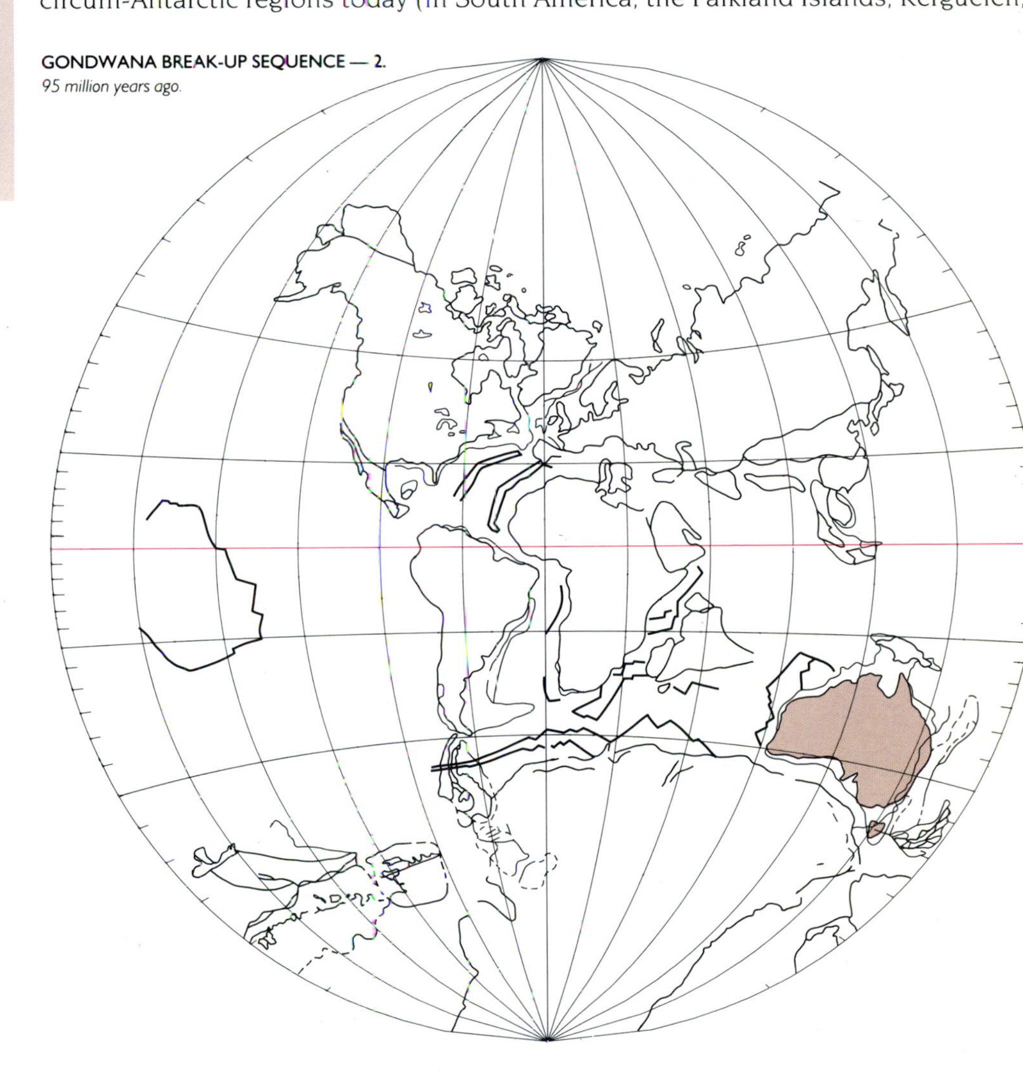

284

285

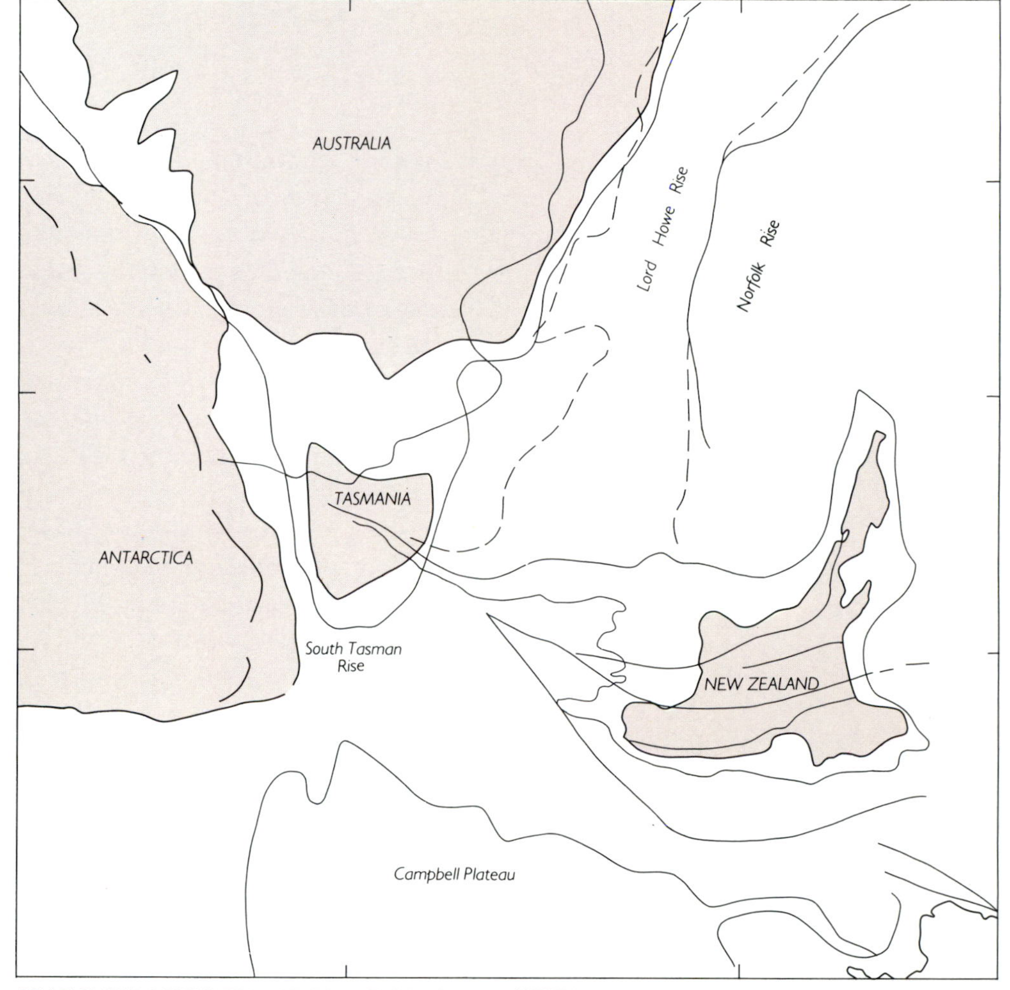

RELATIONSHIP OF NEW ZEALAND TO ANTARCTICA AND AUSTRALIA AT 95 MILLION YEARS AGO.
Broken lines mark linear fractures along which fragments of Gondwana later drifted away from the Australian continental block.

GONDWANA BREAK-UP SEQUENCE — 3.
59 million years ago.

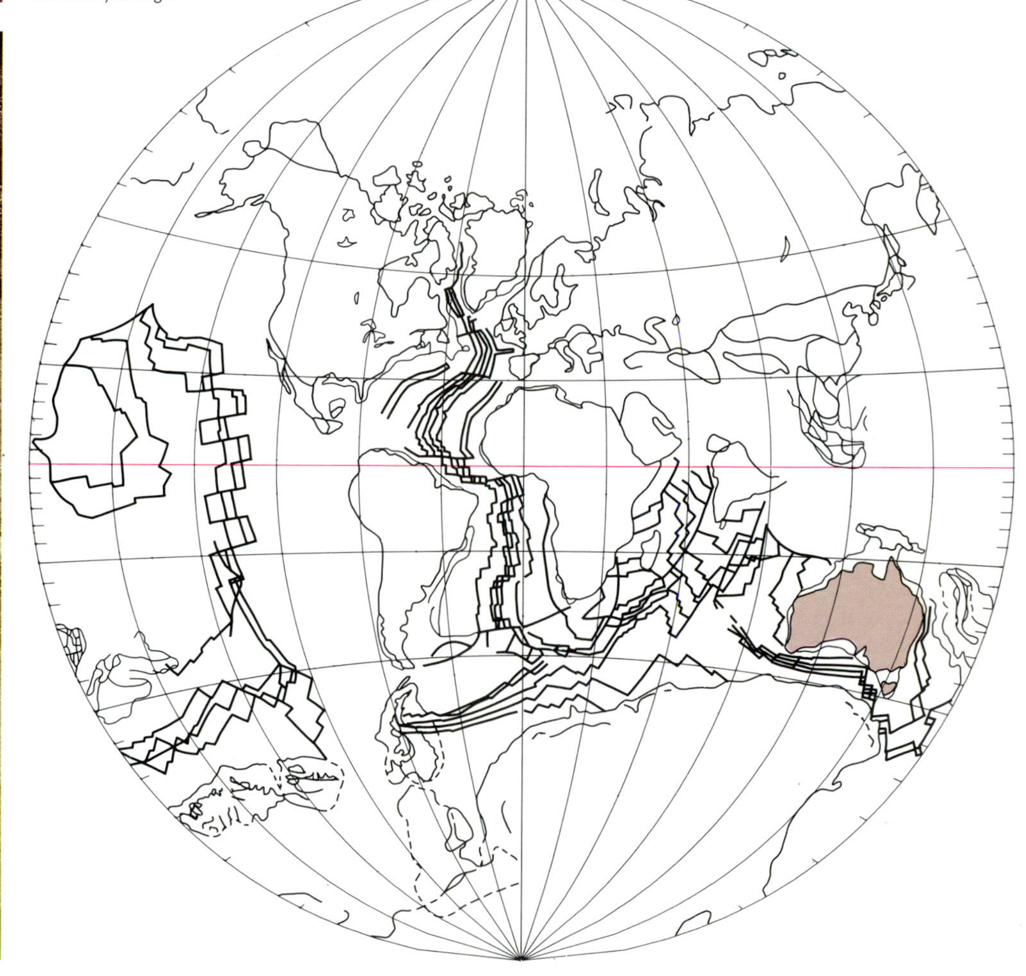

284 *Rissikia sp. (left), a Conifer and Rienitsia variabilis (right) from the Koonwarra fish beds.*

285 *Sphenopteris warragulensis, a delicate thin-leaved Fern from the Koonwarra Fish Beds. The climate was cool with cold winters, resulting in icing of the lake in which the Koonwarra fossils were deposited and winter-kill of Fish. (× 2.0)*

Tasmania, parts of south-eastern Australia, and in New Zealand), is dependent on high, non-seasonal rainfall to attain its fullest expression.

New Zealand's high latitude situation and its weather patterns remain largely suited to the survival of the flora. Its present-day limited extent is due partly to the activities of Man and partly to the effects of volcanic activity which is a feature of New Zealand. Not only does volcanic eruption destroy vegetation by covering areas with ash and pumice as well as igneous rock, but fires are started in adjacent vegetation — and Rainforest has little chance of regeneration after repeated fires.

The similarity of the composition of the present-day New Zealand flora of trees and shrubs to the fossil flora of the Tertiary in New South Wales is remarkable. Fossil FAGACEAE, LAURACEAE, PIPERACEAE, ELAEOCARPACEAE, ARALIACEAE, MYRTACEAE and PROTEACEAE correspond to modern components of the New Zealand flora.

PROTEACEAE are an inconspicuous component in New Zealand's modern flora, in contrast to their dominance in Australia where the family is largely represented by sclerophyll genera. The Australian sclerophyll is an adaptation to poor soils. New Zealand has no comparable sclerophyll, and its soils are comparatively nutrient-rich, being renewed by volcanic activity and subjected to less harsh climatic conditions.

New Zealand has a complement of primitive Angiosperm families including WINTERACEAE and CHLORANTHACEAE. It has a high rate of endemism, as would be expected since evolution of the flora has been in virtual isolation since Palaeocene times.

Separation of landmasses was globally at its most extreme in Late Cretaceous and Early Tertiary times. Continents had not yet formed their new links. However, Australia was still not separated from Antarctica despite the production of new sea floor between them. The South Tasman Rise provided a migration route until between 50 and 45 million years ago when the Rise was breached and Australia's northward movement as the island-continent began.

OTOZAMITES BENGALENSIS

286 *Fronds of Otozamites bengalensis, a Cycadophyte (belonging to the extinct order of Bennettitales) which were abundant in Jurassic and Cretaceous times. They were hardy, drought resistant plants, with thick cuticles, and their abundance in the Northern Territory Early Cretaceous flora probably shows warm and occasionally aridity-affected climatic conditions. From Bauhinia Downs, Northern Territory. Age about 125 million years. (× 1.9)*

286

CHAPTER 9

AUSTRALIA BECOMES THE ISLAND-CONTINENT, WITH A VEGETATION DOMINATED BY THE FLOWERING PLANTS

THE TERTIARY PERIOD

FROM 66.4 TO 1.6 MILLION YEARS AGO

During the Early Tertiary, Australia's last Gondwanan links were severed. The sea floor spreading which had formed the Tasman Sea, isolating New Zealand, continued northwards to form the Coral Sea Basin.

Separation from Antarctica — where sea floor spreading had started in the Late Cretaceous, with the South Tasman Rise maintaining a connection — speeded up. The Rise was breached and ocean waters entered the rift from the west. Australia became the island-continent, moving steadily northwards into increased isolation with its living cargo of plants and animals. This heritage, its quota of Gondwanan life forms, was to have no significant contamination by invasions from other lands for about 30 million years. Evolution within Australia during that time was from the Gondwanan stock, and the individuality of the modern flora and fauna is directly attributable to this fact.

Dryandra praemorsa, a Western Australian member of the PROTEACEAE. The genus is confined to Western Australia, having developed there in the last 5 million years in the isolation which results from the separation of the Mediterranean climatic region from the rest of Australia by intervening desert and semidesert. (Densey Clyne)

THE TERTIARY TIME COLUMN

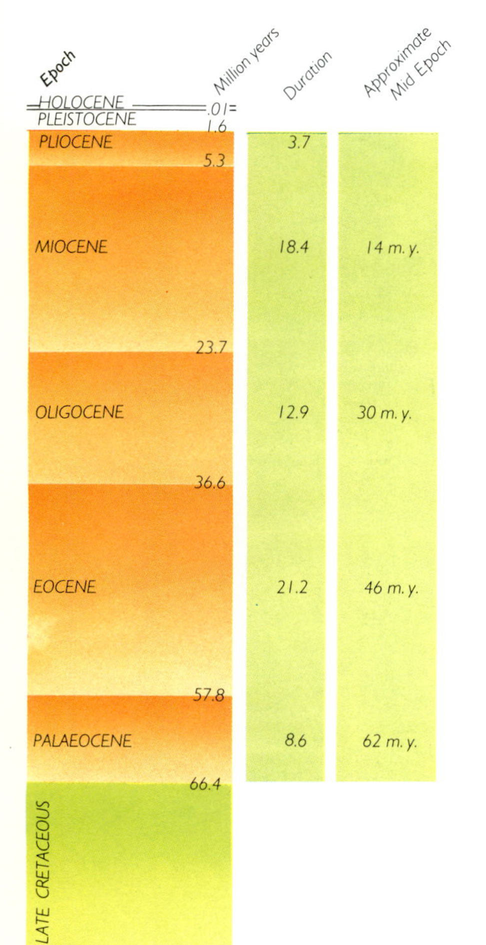

Epoch
Million years
Duration
Approximate Mid Epoch
HOLOCENE
PLEISTOCENE
.01
1.6
PLIOCENE
3.7
5.3
MIOCENE
18.4
14 m.y.
23.7
OLIGOCENE
12.9
30 m.y.
36.6
EOCENE
21.2
46 m.y.
57.8
PALAEOCENE
8.6
62 m.y.
66.4
LATE CRETACEOUS

When the last links with Antarctica were severed, the island-continent embarked on its northward journey with its complement of living creatures. Increasingly isolated, the cargo of this giant ark lived out their lives, and underwent adaptation and evolution with succeeding generations for perhaps 30 million years without any significant input of extra-Australian stock (by which time proximity of the South-East Asian islands allowed for some mixing).

The development of the seaway between Australia and Antarctica did not affect the climate at first. The South Tasman Rise remained as an effective connection until the Late Eocene, about 40 million years ago. It prevented the development of a circum-polar current. Warm equatorial currents travelled to high latitudes along the eastern margins of the southern continents, allowing a transfer of heat from low latitudes. Sea temperatures were probably about 18-20°C at latitudes higher than 60°S during the Early Tertiary. The Southern Ocean was warm and Antarctica was unglaciated. Australia's climate was warm and wet with little zonation. The polar high pressure system was much reduced and a broad zone of westerly winds would have influenced southern Australia. Further north, weak and erratic circulation patterns associated with warm sea temperatures would have resulted in deep inland penetration of rain-bearing winds. 108 105 106

Broad-leaved Rainforest covered much of the continent. Similar Closed Forest grew in Antarctica and in South America. 109 110

As Australia moved further away to the north, and the South Tasman Rise was gradually breached, conditions changed. The Drake Passage formed between South America and Antarctica at the same time. A circum-polar circulation developed and this blocked the equatorial currents, preventing them from approaching high latitudes.

Temperature gradients increased between the Equator and the South Pole, and new weather patterns were gradually established. Westerly wind systems moved further north. The Antarctic continent cooled, the cooling increased circulation, and glaciers appeared on its mountains. By the Late Eocene the vegetation of the Antarctic showed low diversity, evidencing deterioration of conditions for the flora. The low-diversity flora persisted into the Late Oligocene along the northern margin of Antarctica, after which time the cold was too intense for its survival.

Changing and fluctuating weather patterns affected the Australian vegetation from Mid Eocene times onwards. By then the Australia-Papua New Guinea plate had collided with Irian Jaya, starting the deformation of the region.

In the Mid to Late Eocene the seas were still warm and the climate humid. Rainforest was evolving and diversifying into different climatically controlled types dependent on local conditions, and zonation into climatic belts became more evident. The rainfall variations had a greater effect on the Closed Forest types than did the temperature, as is the case today. It was hotter in the west of the continent than in the east. In the Late Eocene to the Early Miocene there was increased seasonality of rainfall and periods of relative dryness.

By the Mid Miocene, about 14 million years ago, the Antarctic ice sheet had reached its greatest dimensions. Atmospheric circulation increased in intensity, probably exceeding that of the present. In parts of Australia levels of precipitation would have fallen and the first onset of aridity may have occurred as a result of the lowering of temperatures and an increase in dry anti-cyclonic circulation. In the Late Miocene worldwide cooling, with much ocean water tied up in ice caps, resulted in a major drop in sea level. When high rainfall episodes occurred, as part of the fluctuating unstable weather patterns of these times of change, river valleys were cut very deep because of the lowered sea level.

287 Bilbies (Macrotis lagotis), accomplished desert dwellers, living under "spinifex" clumps. (Jim Frazier)

287

AUSTRALIA AS AN ARK CARRYING ITS GONDWANAN HERITAGE INTO THE FUTURE

Just as the plants show our Gondwanan origins through the presence of relict Rainforests with their many primitive Angiosperms, and the Southern Conifers, Cycads and Tree-ferns of ancient lineage, so the fauna shows its Gondwanan links.

At the time of separation from other Gondwanan lands, Marsupial and Monotreme ancestors were present. The earliest fossil now recognised as being of mammalian origin is a jaw 100 million years old. Today the only living Marsupials outside Australia are found in South America. This disjunct distribution, with fossils known from the once-linking Antarctica, is the same sort of distribution pattern as is shown by many plant families of Gondwanan origin.

While Australia failed to receive a significant quota of Placental Mammals, which are found in great variety in other continents, the country's fauna has compensated to some degree by evolving into functional equivalents of the different types. Thus we find fossil or living species with ecological and physiological characteristics of Shrews, a Mole, a Dog (the Thylacine), large and small Cats, an Anteater, Flying Squirrels, Antelopes and others. We have, in effect, a parallel but essentially different universe of Mammals: in Australia, Marsupial; in the rest of the world, Placental.

Reptiles of all shapes and sizes were probably living in Australia at the time of separation, and large Monitors are regarded as Gondwanan. The Dinosaurs had been extinct for about 20 million years, having died out at the end of the Cretaceous.

The earliest feather found in Australia, indicating the presence of Birds, occurs in Cretaceous rocks of Victoria. Fossil Penguins and flightless running birds about 50 million years old show that such birds were in the "Australian Ark". The large running birds are Gondwanan — Ostriches in Africa, Rheas in South America, Moas (now extinct) and Kiwis in New Zealand, Emus in Australia, and Cassowaries in Australia and New Guinea. Ostriches were also once present in Europe and Asia, and there were "Elephant Birds" in Madagascar (now extinct).

The Parrots are another Gondwanan bird group, characterising Australia and South America and, to a lesser extent, Africa. Early examples may have had a significant role in the dispersal of Rainforest trees, eating the fruits and distributing the seeds. The Frogmouth Owls of Australia are closely related to the Potoos of South America and share an ancestor.

Some ornithologists suggest that many Australian perching birds, or Passerines, have a common Gondwanan ancestry in a "Crow Ancestor". From this Crow-like bird, all the Flycatchers, Warblers, Babblers, Robins, Wrens, Bower-birds, Birds of Paradise, Magpies, Wood Swallows and the like are thought to have evolved. All these groups of Birds look like their counterparts in Asia and other parts of the world, but modern research on their DNA suggests that the resemblances are the product of convergent evolution. Occupation of similar niches has resulted in similar adaptations and their relationships are Gondwanan rather than cosmopolitan.

The evolution of Birds in Australia was rapid, paralleling the evolution of proteaceous plants and others of the Sclerophyll. There is little doubt that there were already many Birds of many different types and habits by the time Australia started to travel north.

Although most Amphibians have widespread distribution and their places of origin are unknown, there is a group of Southern Frogs, the LEPTODACTYLIDAE, which are regarded as Gondwanan. Turtles, too, occupied the freshwater lakes and streams, and the Side-necked variety is Gondwanan.

The "Australian Ark" also carried a diverse and no doubt immense population of Invertebrates, just as it does today. With the evolution of flowers, rapid evolution of some Insects was occurring, Butterflies in particular. It is a peculiarity of Australia that showy members of many families are Bird-pollinated and that Bees never had as important a function here as in other parts of the world. 123

In the Early Pliocene the Earth warmed and partial melting of the ice caps resulted in a rise of sea level about 4 million years ago. The collision of the Australian Plate with Timor caused the deformation of this region of South-East Asia. There was another cooling episode in the Late Pliocene as the glacial-interglacial fluctuation pattern developed.

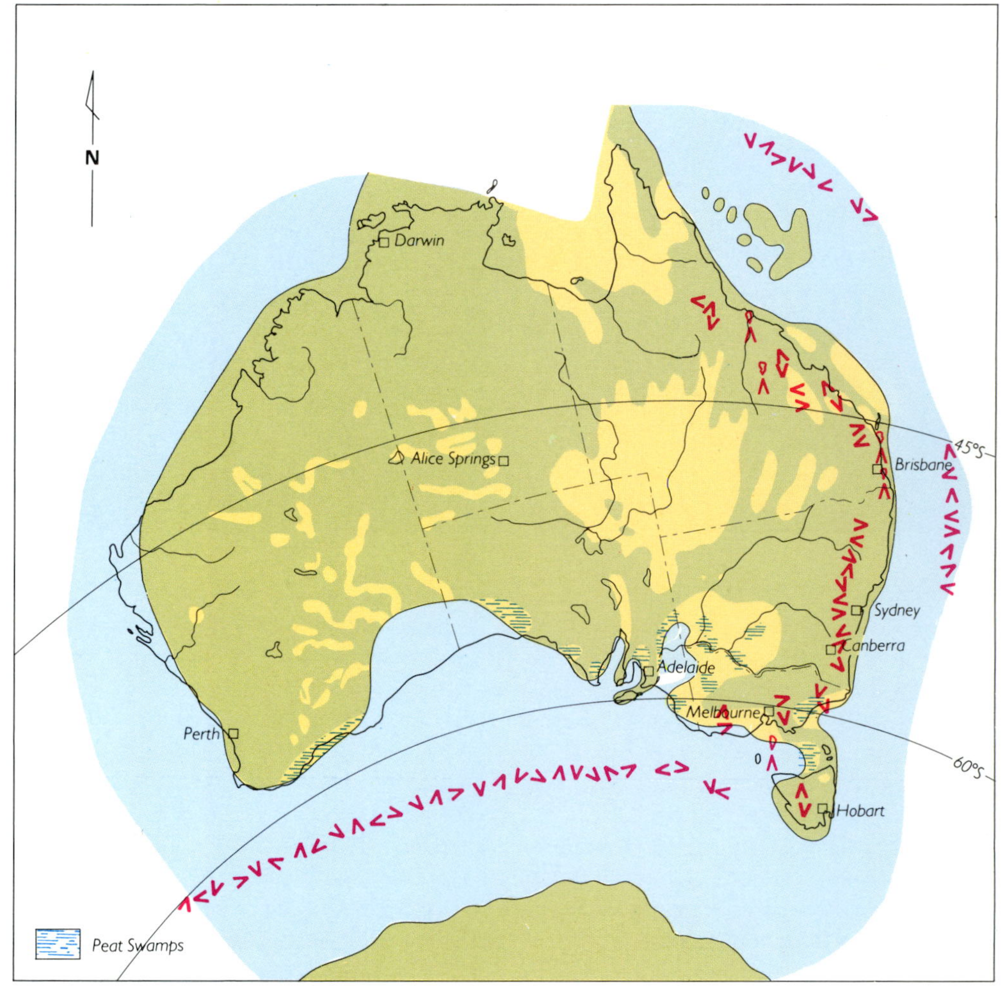

PALAEOGEOGRAPHY OF THE EARLY TERTIARY.

High rainfall meant rapid erosion of rocks, and active drainage systems over most of the continent spread the sediments. Coal swamp conditions existed in parts of the south-east. On the east coast of Queensland oil-shales were formed in lakes north and south of Rockhampton. Warm-temperate Rainforest conditions prevailed over most of the continent, penetrating into the Centre where Broad-leaved Closed Forests existed even at Alice Springs.

Basaltic lavas erupted along the Great Divide and there were submarine flows under the Southern Ocean, the Tasman Sea and the Coral Sea. This igneous activity was related to the rifting processes of the Gondwana break-up. In places along the eastern side of the continent, lavas flowed out over gemstone and mineral-bearing gravels, preserving them from erosion and forming "deep leads".

Australia continued drifting north at 6 centimetres per year. At first the sea flooded the southern edge of the continent. Carbonate sediments were laid down in the Nullarbor Plain incursion. Carbonate deposition also continued in most of the offshore shelf areas of the west and south, through Bass Strait, and also in the Great Barrier Reef area where it formed the precursors of the present reef system.

There appear to have been cycles of deep weathering in much of inland Australia throughout the Tertiary Period. These cycles relate to intermittent high rainfall

288 Landscape at Somerset West, Cape, South Africa, in the Mediterranean climate region, with Leucospermum type PROTEACEAE *(very like Waratahs in appearance) and* RESTIONACEAE, ERICACEAE, *etc., in a "fynbos" community, which is equivalent to Australia's sclerophyll of heathlands. The similarity of the Cape Flora and that of the Western Australian Mediterranean climatic area is striking. (Mary E. White)*

THE MEDITERRANEAN CLIMATIC ZONE OF WESTERN AUSTRALIA

Mediterranean climates develop only on the western sides of continents between sub-tropical deserts and temperate regions, onshore from cold-water currents which are dependent on the existence of polar ice caps. As the glaciation of Antarctica started about 25 million years ago, such climates may have existed in the Southern Hemisphere in suitable locations well before the Pliocene. But even if this were the case, these climates would have post-dated the origin of sclerophyllous plants in Australia. It was not the Mediterranean climate which induced the sclerophylly as some experts had surmised before information was available from the Fossil Record to indicate otherwise.

Western Australia had been effectively isolated from eastern Australia by an inland sea in the Early Cretaceous. Later, in the Early Tertiary, it was again partly isolated in its southern regions by marine incursions into the Nullarbor Plain. This isolation accounts for the high rate of endemism in its flora (about 80 per cent). The establishment of Mediterranean climatic conditions about 5 million years ago was followed by diversification of sclerophyllous genera in southern Western Australia and evolution of many more endemics. Many of the plants which have evolved under those conditions of winter rainfall still show growth-peaks in late summer, indicating that their ancestors had evolved under summer rainfall regimes.

The development of the characteristic Mediterranean sclerophyll flora of Western Australia and the similar Cape Flora (known as "Fynbos") of South Africa are examples of parallel evolution under similar climatic conditions on soils with low nutrient status. From the original Gondwanan stock of early Angiosperms, related plants occupy the same niches and similar adaptations have fitted them for their roles. Thus they have the same general appearance and are dominated by the same families.

288

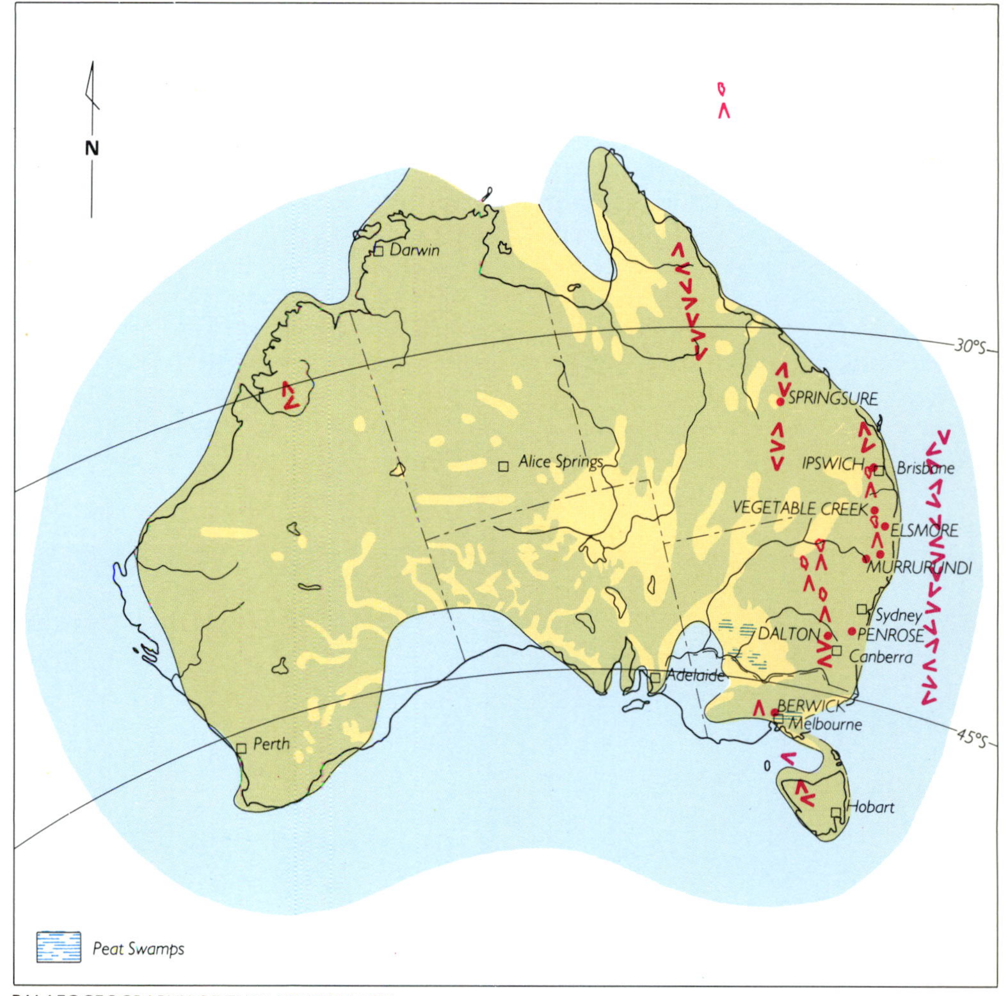

PALAEOGEOGRAPHY OF THE LATE TERTIARY.

episodes, when ephemeral rivers supplied sediment to alluvial tracts and lakes. Towards the end of the Period, the phases of deep weathering were further reinforced, leading to the formation of deposits of bauxite and of opal. Leaching changed the near-surface expression of many mineral deposits. Uranium, from the weathering of granite and other rocks, accumulated in river and lake deposits in Western Australia and South Australia.

Basaltic eruptions continued along the Great Divide and formed guyots in the Tasman Sea. In the Kimberleys of Western Australia, south and west of Wyndham, igneous rocks from deep in the Earth's crust and mantle were intruded into Late Tertiary rocks about 20 million years ago and several of them formed volcanic "pipes" which contain diamonds.

Tasmania was part of the Australian mainland at least until the Oligocene or the Miocene when Bass Strait was formed. By then the Australian flora was well defined with a Rainforest element as well as the Sclerophyll.

The seaway between Tasmania and the mainland has been open and shut at least eight times since, due to fluctuations in sea level. As a result the island-State has no endemic families and only about 20 per cent of its species are endemic. Because it has maintained a climate similar to that in which the Antarctic Beech-Southern Conifer flora arose, it (like New Zealand) has many representatives of Gondwanan families today.

AUSTRALIA'S ARID CENTRE

The hot and dry Centre of the Australian continent has not always been arid. In Early Tertiary times it supported Rainforest. Changing and fluctuating weather patterns from the Mid Miocene onwards resulted in the establishment of seasonal rainfall regimes with dry periods. There was a gradual changeover from Closed Forest to open woodlands, then to dry sclerophyll scrublands, and later to grasslands.

About 15 million years ago, in the Mid Miocene, broad-leaved forest covered the area now occupied by the Simpson Desert. The circum-Antarctic Gondwanan flora comprising these forests probably had an admixture of sub-tropical and tropical plants which had entered Australia by the Indian-Asian route. For about 30 million years the floras derived from the northerly and southerly migration routes had been able to mix while the climate was uniformly suitable.

With permanent freshwater lakes, conditions were ideal for a diverse and thriving fauna. Giant Koala-like Possums, many types of Ringtail Possums, small insectivorous Marsupials (like the South American

289

290

291

Opossum), primitive Bandicoots, sheep-sized Marsupial Browsers (Diprotodontids) and primitive Kangaroos like modern Tree-kangaroos have all left a fossil record. There were flocks of Flamingoes (which share an ancestor with Ducks) and giant Flightless Birds (Ratites). Crocodiles, Turtles and freshwater Dolphins lived in both the lakes and the rivers.

The gradual contraction of areas in which the lush conditions prevailed resulted in the extinction or evolutionary adaptation of the animals and plants which remained in the Centre. Today it is the home of some of the world's most accomplished desert dwellers, like the Greater Bilby, the Marsupial

289 "Spinifex", Saltbush and shrubby Acacias, and flat-topped mesas. (Densey Clyne)

290 Marsupial Moles (Notoryctes typhlops), which live in the desert, are an example of convergent evolution. They are adapted to their subterranean, burrowing existence in the same ways as Moles (Placental) are in Europe and elsewhere. (Jim Frazier)

291 An artesian waterhole, William Creek, northern South Australia. Saltbush, hummock and tussock grasses and salt-tolerant herbs in the wet region, Mulga scrub on the hillside. (Glen Carruthers)

Mole and the many species of Hopping Mice. Australia's desert plants, derived from dry sclerophyll, are predominantly not succulent like their counterparts in Africa, nor are they spiny and water-storing like the Cactus-type plants of South American deserts.

As the Centre dried, small pockets remained where the lush conditions persisted locally, allowing animals and plants to survive virtually unchanged. As parts of north-eastern Queensland and New Guinea attained altitudes at which suitable habitats existed, relict flora and fauna refugia were created. The Mammals living in these highland areas today bear a striking resemblance to the extinct types of Mammal found as fossils in the 15-million-year-old fossil deposits of present-day Central Australia.

River systems in arid Western Australia have been inactive since the Mid Miocene and are now reduced to chains of salt lakes. In the Lake Eyre Basin climatic changes have been well documented, showing an alternation of humid and dry periods throughout the Tertiary, with the present degree of aridity not appearing until the Quaternary, about 2-2½ million years ago. Similarly, in the Lake Frome area the climate of the Miocene was warm with high rainfall and seasonal aridity. Arid and pluvial climates alternated in the Late Tertiary.

The aridity of the Australian continent reached its present proportions in Late Pliocene to Early Pleistocene times, about 2 million years ago.

With relative aridity occurring in some areas at least since the Mid Eocene, and with the possibility that semi-desert to desert existed in localised areas (due to topography and specialised local conditions), there was pressure for the evolution of arid-adaptations in the flora. In the same way that sclerophyll adaptations resulted from the contraction of Rainforests as the climate became unsuited to Closed Forest, so the Arid Zone flora evolved in the drier pockets and zones increasingly affected by aridity. Its components came from the dry sclerophyll vegetation of adjacent areas. Sclerophyllous adaptations already suited the plants to water conservation. Their small, often leathery and hard leaves, small growth habits, mechanisms for seed preservation under adverse conditions, root adaptations and other features made the transition easy.

An interesting side effect of the adaptation of sclerophyll to arid environments is seen in the changes in pollination mechanisms. For example, in the PROTEACEAE, Bird and non-flying Mammal pollination has been an important part of the evolutionary history. Under arid conditions there has been a dramatic decrease in pollination by these agents and a corresponding increase in wind and Bee pollination.

The appearance of *Eucalyptus* pollen in the Fossil Record for the first time in Early Oligocene times (34 million years ago), and then of *Acacia* in the late Oligocene (25 million years ago), shows the changes to sclerophyll vegetation over most of the continent as Rainforest contracted. The abundance of grass pollen 5 million years ago in the Early Pliocene shows the transition to open grassland in some areas and to grasses in open woodland generally.

The ASTERACEAE, or Daisy family, appears in the Pollen Record in the Mid Miocene and increases rapidly. This family provides many of the ephemerals which bloom after rain in the deserts. The family is cosmopolitan and not part of the Gondwanan heritage. It is dispersed by wind-blown seeds, and the adaptation of many of its members to sand-dune dwelling on the coastal fringes of continents meant easy spreading between landmasses as well as suitability to desert conditions after rain.

292

292 Ephemerals after rain, all showing sclerophylly — leaves are reduced, or spiny, or succulent. (Densey Clyne)

WATERLILIES AMONG THE EARLIEST ANGIOSPERMS

In areas close to the point of origin of the Flowering Plants — in North Africa, northern South America, Europe and North America — the very early Angiosperm pollen types are found in Early Cretaceous sediments. They are of general types which show affinities with *Magnolia*, Lilies and Laurels, but cannot be assigned to any living genera. Among the first to be assignable to a family is Waterlily pollen (Magnolioid), recorded from the Early Cretaceous in Europe.

The Australian Pollen Record for Angiosperms starts in the Late Cretaceous, by which time diversification had taken place and the pollen types of different families were recognisable, notably *Ilex*, WINTERACEAE and primitive PROTEACEAE.

293

293 Waterlilies in a Northern Territory Lagoon. (Jim Frazier)

AUSTRALIA'S FOSSIL POLLEN RECORD

105 111

FIRST APPEARANCE OF POLLEN TYPES IN SOUTH-EASTERN AUSTRALIA

Gymnospermous pollens extend from the middle of the Mesozoic through the Tertiary in the Pollen Record.

The first recognisable Angiosperm pollen is of *Ilex*, a Holly, which occurs in the Late Cretaceous. It is closely followed by Antarctic Beech and proteaceous types.

The MYRTACEAE, perhaps the most important family today, first appears in the Palaeocene.

Taxa appear almost continuously throughout the Tertiary, with more herbaceous taxa appearing in the Late Tertiary.

In the Palaeocene, 66-57 million years ago: the Gymnosperm group consisted of ten different pollen types and was dominant. The proteaceous group of up to 23 different pollen types is also abundant. Many of the proteaceous types become extinct during the Tertiary. MYRTACEAE is present but not abundant, as is Antarctic Beech. Other Angiosperms continue to diversify.

In the Eocene, 57-36 million years ago: the early assemblages are like those of the Palaeocene. In the Mid Eocene, about 40 million years ago, Antarctic Beech suddenly becomes dominant. It is a heavy pollen producer, and is dispersed by wind, so it is always over-represented in samples. At the time it becomes dominant many of the proteaceous types and some other Angiosperm types become extinct.

In the Miocene, from 23 to 5.3 million years ago: the early pollen floras are the same as in the Oligocene. In the Mid Miocene, Antarctic Beech decreases and MYRTACEAE becomes dominant with eight different types. The "Eucalypt pollen type" which may be found in other genera besides *Eucalyptus* is present but not abundant, and most of the pollens are of *Tristania*, *Backhousia*, *Baeckea*, *Syzigium* and *Acmena* (and probably other genera).

In the Pliocene, from 5.3 to 1.6 million years ago: MYRTACEAE is dominant with small amounts of Antarctic Beech and other early Tertiary forms. Grass pollen and ASTERACEAE (Daisy family) are usually present, but in low frequencies.

In the Pleistocene: Antarctic Beech and most of the Gymnosperms have disappeared from most of the continent. MYRTACEAE is dominant, and grasses and the Daisy family have increased greatly.

(*After* Martin, 111)

The Angiosperm Pollen Record in Australia is only partially known, with a strong bias to south-eastern Australia. It may be many years before a comprehensive picture is available for the whole continent. Meanwhile, it is unwise to apply the information from the south-east to the whole. As climatic zonation increased throughout the Cainozoic, diverse floral elements evolved.

Some major groups of plants, notably the Laurales, have pollen which does not survive fossilisation. As this is one of the earliest Angiosperm types, and as many of the first Flowering Plants in Australia were probably Lauralian, this lack of fossilisation represents a significant problem in interpreting the flora from the Pollen Record. It is necessary, also, to remember that wind-blown pollen will be over-represented in any sample. Further, only in climatic zones with moderate rainfall are conditions likely to favour any fossilisation at all. Though the Pollen Record gives a much better account of the flora than does the macrofossil record, the information is still far from complete.

Because of the vast numbers of families and genera of Angiosperms in Australia, only a small selection can be discussed. The plants chosen show the Gondwanan links of modern and ancient floras. Present-day distribution patterns of some modern plants, which may not have a fossil record, also show the Gondwanan origin.

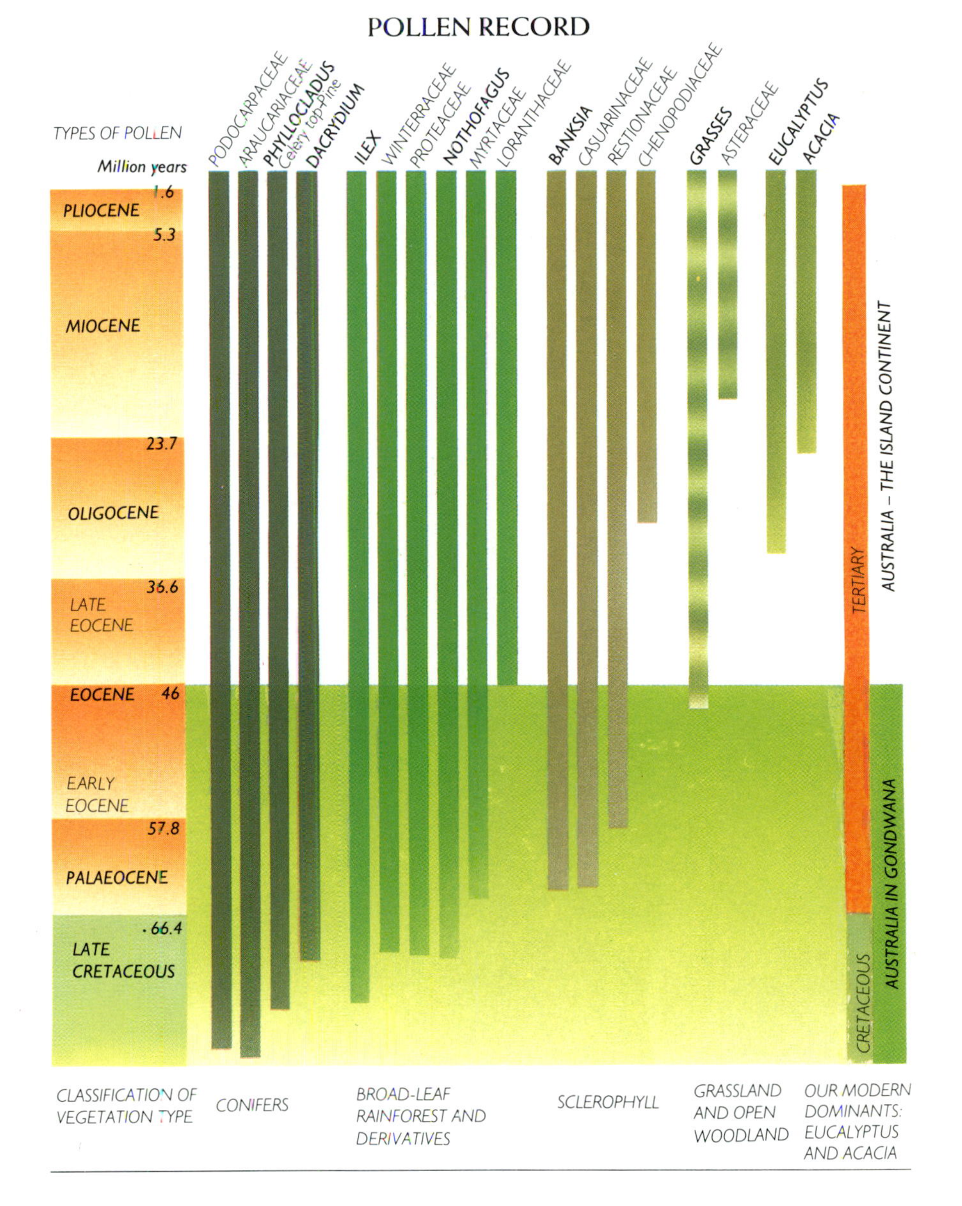

294 Bunya Pine *(Araucaria bidwillii)* was once abundant and formed forests of considerable extent. Now small relict stands and isolated specimens remain. The Aborigines used the Bunya as a food source, planning their walkabouts to coincide with the ripening of the enormous female cones, larger than a football, which are a source of high protein seeds. *(Densey Clyne)*

295 A Hoop Pine, *Araucaria cunninghamii*, which is now also found locally abundant in a few refugia, having been widespread and abundant in the past. Hoop Pines form a stage in the development of Rainforest, occurring in the mixed communities fringing the ancient broad-leaf Closed Forest. *(Densey Clyne)*

Australia has no record of the earliest Angiosperm pollens of the Early to Mid Cretaceous. Remote from the region of West Gondwana where the Flowering Plants originated, the first Angiosperm pollen so far recorded in this country is *Ilex*, a member of AQUIFOLIACEAE (Holly family) from the Late Cretaceous. The genus was then distributed worldwide with many species. (*Ilex* has seeds which are distributed by birds and its embryos undergo continuing development until germination, which may take from 2 to 8 years — strategies which must be an aid to wide distribution.) 112

Ilex is prominent in the Pollen Record until the Miocene, after which it declines. Today there is one relict species, *Ilex arnhemensis*, occurring in wet Rainforest in the Northern Territory, the Atherton Tableland of Queensland, and northern Western Australia. It also occurs as a lowland swamp species in New Guinea, where there are ten other species. There, conditions have remained suitable, providing a refuge for these plants which spread initially while Australia was part of Gondwana.

The Southern Conifers still dominate the Pollen Record in the Late Cretaceous, being only gradually ousted by the Angiosperms. They are plants with low dispersibility, requiring continuous land for migration.

294

295

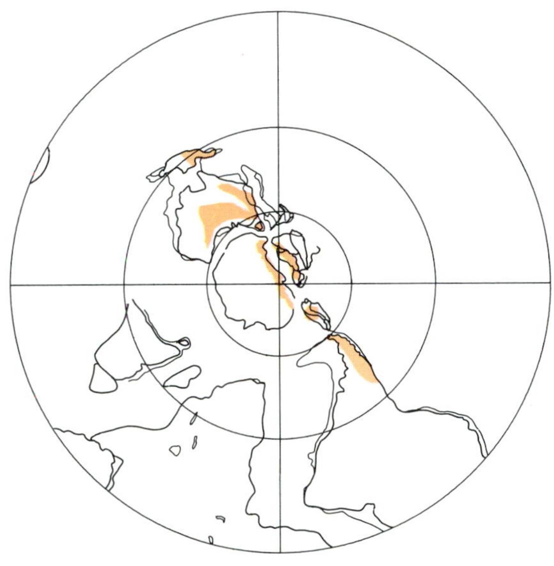
DISTRIBUTION OF *NOTHOFAGUS* IN GONDWANA

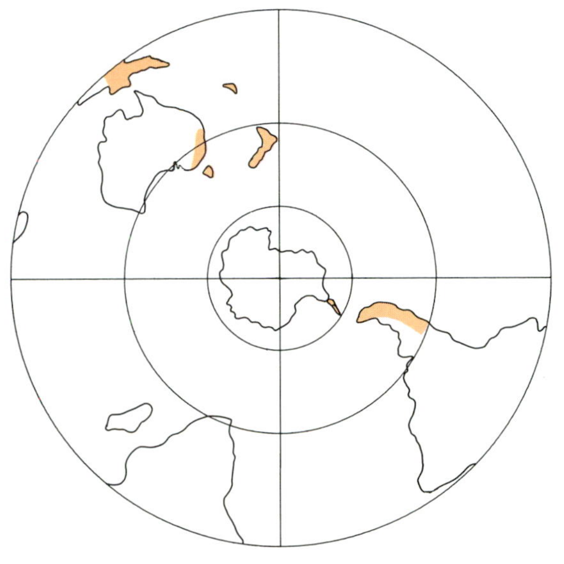
MODERN DISTRIBUTION OF *NOTHOFAGUS*
After G.R. Stevens, 1980, with permission.

FAMILY FAGACEAE, GENUS *NOTHOFAGUS* — THE ANTARCTIC BEECHES

113

Nothofagus appears in the Pollen Record in the Late Cretaceous, after the appearance of *Ilex*. At the same time, primitive PROTEACEAE and WINTERACEAE made their first appearance. Pollen of Antarctic Beeches remained abundant until the Mid Oligocene, then declined, and there is an increase in *Araucaria* pollen to compensate. *Araucaria* (Hoop Pine), associated with Podocarps, seems to represent a stage between wet Rainforest and Sclerophyll. Pockets of Hoop Pine forest in Queensland today are relics of the once widespread communities of the Mid Tertiary.

Nothofagus today is found in New Guinea, New Caledonia, New Zealand, southern South America and in Rainforests of south-eastern Australia. It is a plant with low powers of dispersion, requiring continuous land for its migration. It seeds are not wind-dispersed and they are rapidly killed by saltwater.

Nothofagus is absent from Africa, having no fossil pollen record there and no living examples. It must have evolved after links between Africa and South America had been severed, and if it evolved in South America its spread must have been from there to Antarctica to Australia. However, its place of origin could have been elsewhere in the Gondwanan landmass. Records are too incomplete to be sure where the oldest record of its pollen is found.

296

296 Nothofagus moorei, a twig of Antarctic Beech, showing the female flowers in cone-like heads. Nothofagus occurs in relict Rainforest in Tasmania and south-eastern mainland Australia, and in the mountains of New Guinea. (Peter Valder)

DISTRIBUTION OF WINTERACEAE

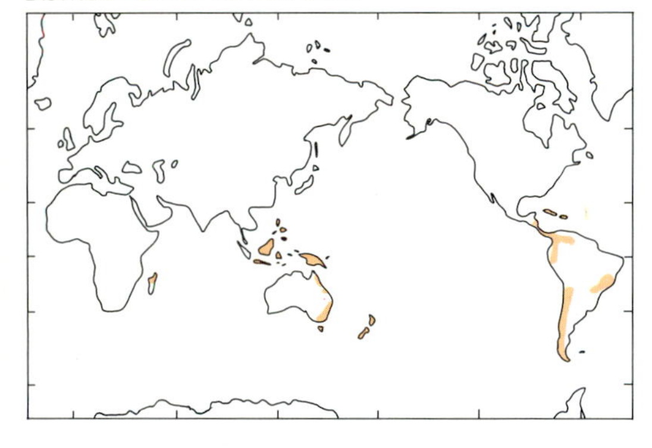

FAMILY WINTERACEAE

WINTERACEAE, one of the most primitive Angiosperm families alive today, has a relict Gondwanan distribution pattern. The genus *Tasmannia* is a member of Australia's Rainforest communities. 114

Fossil pollen of WINTERACEAE has been found in Tertiary sediments (probably Miocene) in South Africa and in Cretaceous sediments in Israel. These finds show that the family was present in other parts of Gondwana in the past. 115 116

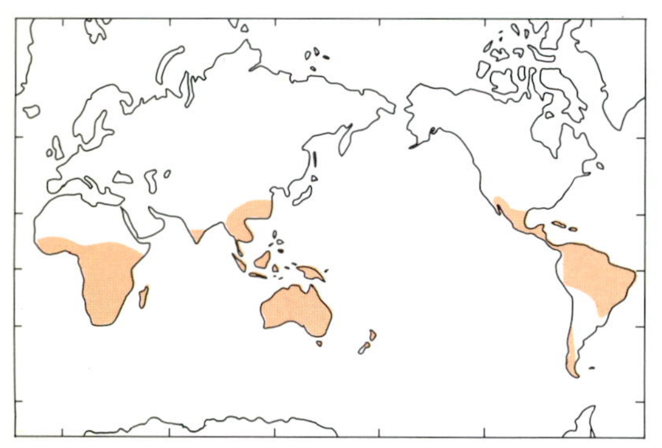

DISTRIBUTION OF PROTEACEAE

297 Flowers of *Banksia ericifolia*. *(Densey Clyne)*

298 *Oreocallis wickhamii*, a Queensland Rainforest tree. *(Peter Valder)*

299 *Embothrium lanceolatum*, from Chile, closely related to the Australian rainforest tree *Oreocallis wickhamii*. *(Peter Valder)*

300 A *Protea* with *Erica* (the heather), in a South African "fynbos" community in the Cape. The ERICACEAE take the place occupied by the EPACRIDACEAE in Australian sclerophyll. *(Mary E. White)*

301 *Leucospermum*, a genus of PROTEACEAE endemic to South Africa, is very similar in appearance to *Telopea*, the Waratah, which is endemic in Australia. *(Mary E. White)*

FAMILY PROTEACEAE, AN ANCIENT GONDWANAN LINK

117

The proteaceous pollen associated with Antarctic Beech and WINTERACEAE in the Late Cretaceous represents the primitive Rainforest component of the family's long history. Today, in the northern Rainforests of Australia, *Oreocallis* grows as a Gondwanan relict and its closest living relatives occur in Chile.

At the time of diversification of the PROTEACEAE into sub-families, tribes and most of the sub-tribes, and their spread amongst the southern landmasses, they were trees or at least sizable woody plants growing in Closed Forests.

The PROTEACEAE plant family is characterised by low dispersibility, requiring continuous land for migration. Its modern distribution is entirely Gondwanan. After its origin in West Gondwana, Africa separated and its proteaceous plants developed in isolation. As a result, there is no modern genus in common with Australia and the Indo-Malayan area. But parallel evolution from a common stock has resulted in suites of very similar plants in Africa and Australia.

118

As proteaceous plants were among the first Angiosperms to appear in the Fossil Record in Australia, they must have been involved in the transitional ecosystems between the ancient floras and the modern-aspect Rainforest flora of the Tertiary. Their fossil record can be interpreted in terms of the evolution of Rainforest.

By the Palaeocene Period, 65 million years ago, the first primitive proteaceous pollen types had given way to recognisable modern generic types and *Banksia*-type

297

298

299

300

301

pollen was abundant. There seems little doubt that the ability of Banksias to colonise barren sand by virtue of "proteoid roots" accounts for the markedly greater abundance of this type of pollen early in the Fossil Record. After the vast marine incursions of the Cretaceous, very large areas would have been denuded of vegetation. In terms of soil, these areas would have been largely silica sand. Because of the flatness of the continent, the pattern would have been one of relatively sudden inundation of large areas followed by rapid re-emergence of large areas of land when the sea level eventually fell. Enormous swamps would have been a feature of landscapes during the process.

Consider the sequence of events during the period when the Flowering Plants were arriving in Australia and diversifying here. While the first Angiosperms were appearing in West Gondwana and radiating out into adjacent areas of the Northern and Southern Hemispheres, under conditions which appear to have been warm and wet, most of Australia was flooded. The climate was cool to cold (as evidenced by ice-rafted dropstones in the Eromanga Sea). The Australian landmass was on the outer edge of Gondwana and already becoming separated from land to its north. As the climate warmed up and the seas retreated, Flowering Plants were beginning to invade the relict Gymnosperm-Fern communities. These communities had carried over from the luxuriant communities of the uniformly warm and wet Jurassic, when they were at their peak worldwide. Climatic changes with increasing cold had put them under stress during the Cretaceous and their ancient conservatism denied them the flexibility required to adapt to changing climate and environments. (The situation was paralleled in the Animal Kingdom by the Dinosaurs, which suffered extinction after being stressed by a change to cold climate which posed difficulties for cold-blooded creatures.)

III

The Pollen Record for the Late Cretaceous shows a high proportion of Conifer pollen and small amounts of pollen from primitive PROTEACEAE and WINTERACEAE. During the Palaeocene, as the climate warmed up and became uniform over the continent, with high rainfall, the Broad-leaved Rainforest community was evolving and would become the dominant vegetation type in the Eocene.

To think of all the early Angiosperms as being Rainforest trees and shrubs and all sclerophyllous plants as derivatives from Rainforest ancestors is not logical. Surely Sclerophyll and Rainforest must have co-existed from very early times in Angiosperm history, the proportion of one to the other changing with time and local conditions. The ancient Cycad–Conifer–Ginkgophyte–Fern communities in Australia were largely established on poor soils and presumably had mycorrhizal and other mechanisms to enhance mineral uptake as members of those plant groups do today. They were the first sclerophyll, and the angiospermous sclerophyll which was emerging and forming a succession towards Rainforest on low-nutrient soils would have had proteoid roots (not confined to PROTEACEAE), mycorrhiza and other symbiotic and structural arrangements enabling them to succeed outside the Rainforest ecosystem with its recycled nutrient-rich soils.

In the extensive swamp environment formed as the epicontinental seas retreated, some of the first Flowering Plants would have been sclerophyll, adapted to the low nutrient conditions in swamps. Many would have been extremely salt-tolerant. The early arrival on the scene of RESTIONACEAE, *Banksia*-type pollen, CASUARINACEAE and MYRTACEAE can be seen in part as a reflection of sclerophyllous communities developed in response to swamp environments.

Broad-leaved Rainforest, which clothed much of Australia in the Late Palaeocene and the Eocene, could only have become established on old sea-bed sands as the end-product of a succession from pioneers on dunes and pure silica sand in general, through several stages, each contributing to the environment. Though Rainforest

302

303

302 *Telopea speciossima*, the Waratah, State flower of New South Wales. The genus is confined to Australia. *(Densey Clyne)*

303 *Banksia coccinea. (Densey Clyne)*

304 Casuarinas colonising pure sand on Fraser Island. *(Mary E. White)*

may grow on sand-dunes, as it does on Fraser Island and the Great Sandy Region of Queensland, it does not spring up as an entity on an area of new sand simply because the rainfall and climate are right. Research on Fraser Island has thrown new light on succession and the concept of "climax" forest. Much of the climax model was based on evidence from Europe where landforms are very young and date only to the last ice age. Where landforms are old and soils are depleted by weathering over the ages, the situation is not as simple.

The evolutionary process from sand-dune to Rainforest, as seen on Fraser Island today, presents the most plausible model for the evolution of Tertiary Rainforests. Fraser Island is the largest sand island in the world. It has been built up by the addition of sand to its eastern (outer) edge by ocean currents. It shows a succession of dunes from east to west, increasing in age and mirroring the evolution of vegetation through time as the Rainforest came into dominance. Bearing in mind that grasses, for instance, which colonise the dunes today, did not appear on the scene until much later in the evolutionary sequence, and that modern representatives of groups are not necessarily the same as fossil ones, the concept is that plants similarly adapted to niches formed a succession in the past. Today *Banksia* and *Casuarina* occupy the youngest dunes and Mycorrhizal Fungi form dense masses in the sand, spreading out ahead of the advancing line of vegetation. Dry sclerophyll scrub and woodland follow, and wet sclerophyll with Hoop Pine leads into the Rainforest.

In the special case of Fraser Island, where there is no bedrock anywhere near the surface, the soil-nutrient layer gets progressively deeper with the aging of the dunes. Rainforest has immensely deep root systems to tap this nutrient. On the oldest western dunes the nutrient layer is too deep even for Rainforest roots and the succession is reversed with sclerophyll increasingly dependent on mycorrhiza and other arrangements for nutrient enhancement on poor soils. The reverse succession from Rainforest through increasingly dry sclerophyll is a mirror of the evolution of our modern Sclerophyll when the climate dried out and Rainforest was confined to relictual habitats in a continuing process starting in some areas in Late Eocene times.

304

FAMILY CASUARINACEAE — THE SHE-OAKS, DESERT-OAKS AND RIVER-OAKS

III

The appearance of CASUARINACEAE pollen in the Palaeocene, at the same time as that of Banksias, is further evidence of the establishment of sclerophyll related to the colonising of sand. Today's Casuarinas produce seasonal proteoid roots, have V.A. mycorrhiza, and are adapted for colonising sand-dunes. The further adaptation in some species for swamp living, by production of aerating roots which grow upwards as in Mangroves, is an important evolutionary advance.

The family is now largely Australian, with 45 endemic species. A few species occur in South-East Asia, and one species occurs in both Madagascar and the Mascarenes. A related genus, *Gymnostoma*, is found in New Guinea and New Caledonia.

The family is also interesting botanically, being closely related to the catkin-bearing FAGACEAE and showing some morphological features suggesting affinity with Gymnosperms rather than Angiosperms.

305 A Desert Oak (Casuarina sp.) with "spinifex", in the Red Centre. (Densey Clyne)

305

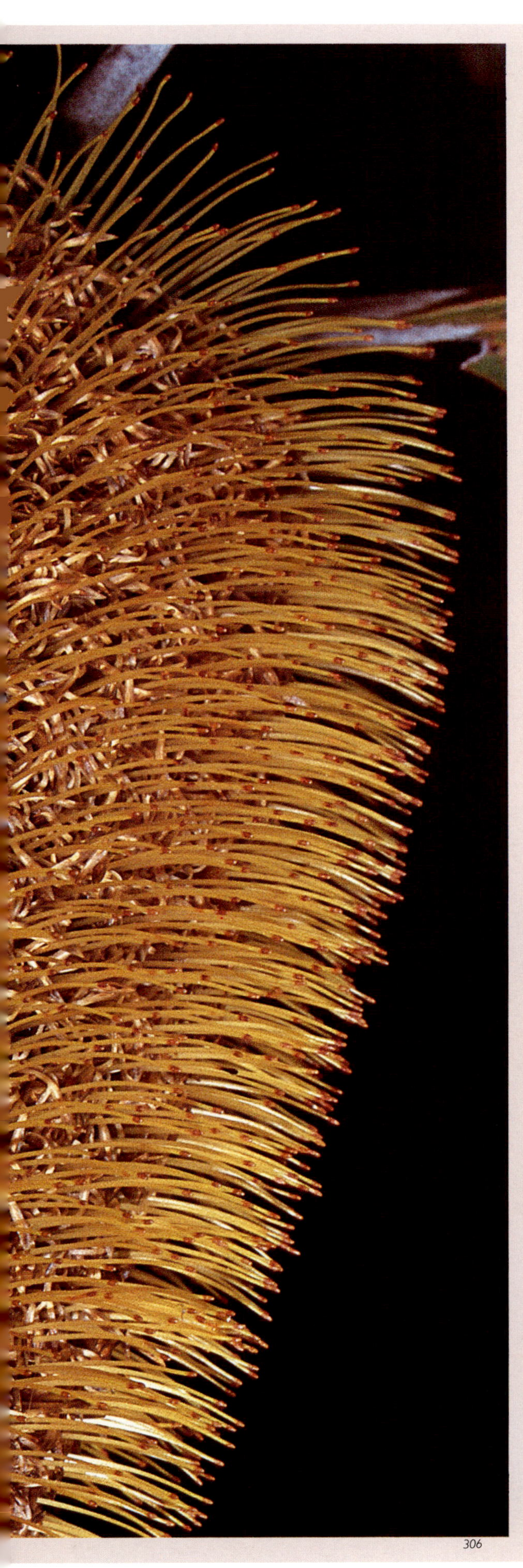

FLOWERS POLLINATED BY BIRDS AND MAMMALS

The parallel evolution of PROTEACEAE and birds of the Honeyeater family, MELIPHAGIDAE (Order Passeriformis), is a fascinating feature of the pollination story in Australia.

The country's 66 species of Honeyeater are the main pollinating agents for *Banksia*, *Dryandra*, *Grevillea*, *Hakea*, *Lambertia*, *Telopea* and other PROTEACEAE. They also feed on flowers of many other families, notably Eucalypts (MYRTACEAE).

Undoubtedly, ancestral Honeyeaters were associated with the first Banksias, evolving with them in the sclerophyll and also playing their part in pollination of early MYRTACEAE and other suitable flowers of trees in the Rainforest. Today's Honeyeaters have brush-tipped tongues for taking up nectar, and most of them supplement their diet with insects. The nectar contains amino-acids and other substances essential for the birds' well-being, as well as the sugars (which are the main attraction). In the wonderful ways of Nature, proteins in the nectar of flowers attended by Honeyeaters which have little else in their diet have higher concentrations than in flowers whose visitors have a high-protein diet by virtue of their more highly developed insectivorous habits.

An example of parallel evolution of a complementary feature is seen in the Rainbow Lorikeet (*Trichoglossus haematodus*) and its allies, which have brush-tipped tongues like the Honeyeaters to fit them for their role as nectar and pollen feeders. They do their share of pollinating, visiting Firewheel Trees (*Stenocarpus*, PROTEACEAE) as well as the flowers of many different forest trees, following the blossoming in much the same way as Bees do.

Another fascinating animal-plant pollination relationship is that which exists between Honey Possums (and also Gliders) and Banksias. These tiny, nocturnal Marsupials transfer pollen on their fur and noses from one plant to another as they eat pollen and nectar. In southern Western Australia there is a Brush-tongued Honey Possum (*Tarsipes*) which has developed the same sort of adaptation to its tongue as have the Honeyeaters and Lorikeets.

Bats, too, act as pollinating agents for some large-flowered species.

Under more arid regimes, sclerophyll plants which are elsewhere Bird-pollinated become Insect-pollinated. Indeed, Insects always play some part in the processes, being attracted by the nectar just as the Birds are.

306 Bat pollinating Banksia flowers. (Jim Frazier)

307 Honeyeater pollinating a Waratah. (Jim Frazier)

306

307

FAMILY MYRTACEAE

The distribution of MYRTACEAE, the family to which the genus *Eucalyptus* belongs, is fundamentally Gondwanan, in spite of considerable tropical distribution and the northern extension of a few groups. 117

The family has two main sections: Myrtoideae, in which the fruits are fleshy; and Leptospermoideae, in which the fruits are dry capsules.

An early radiation of ancestral MYRTACEAE was followed by parallel evolution in the continents as Gondwana broke up.

True Myrtles are the only modern group in the Northern Hemisphere. Africa is poorly represented, having only two genera with fleshy fruits (which have an extra-African distribution as well) and two with dry fruits (which are endemic).

The family is predominantly represented by dry-fruited genera in Australasia where there are 95 genera, 93 of which are confined to the region. Australia itself has 69 genera — 55 are dry-fruit (including 42 endemics) and 14 have fleshy fruits (of which three are endemic). The abundance of the dry-fruit type is related to the sclerophyll habit of most Australian plants. The fleshy-fruit genera are trees and shrubs of sub-tropical and tropical woodland and Closed Forest, and represent the ancient element of the flora.

South America and Central America have 47 genera of the fleshy-fruit group, of which 46 are endemic.

Flowers of MYRTACEAE are mainly pollinated by animal agents — Insects, Birds, Bats and small Mammals — and most of them produce nectar to attract their visitors. There is a widespread tendency towards evolution of "brush blossoms" in which the stamens are the most conspicuous organs.

308

309

310

308 Baeckea grandiflora, another member of MYRTACEAE *with dry fruits (three celled capsules) but with the Tea-tree flower type. (Densey Clyne)*

309 Eremaea fimbriata, a dry-fruited member of the MYRTACEAE. *A dry, grey capsule is seen at the left of the picture. Flowers are brush-blossom type. (Densey Clyne)*

310 Brilliantly coloured, fleshy fruits of Syzygium luehmannii, a member of the Myrtoidae section of MYRTACEAE. *The common name for fleshy fruited examples is Lilly Pilly. (Peter Valder)*

311 Angophora growing on Hawkesbury Sandstone near Sydney. Huge trees often grow out of cracks in the rocks, with their roots following the contours of the rocks and growing into grotesque shapes. (Densey Clyne)

GENUS *EUCALYPTUS*

Eucalyptus pollen first appears in the Fossil Record in the Early Oligocene (about 34 million years ago), some 30 million years after the family MYRTACEAE to which it belongs had made its debut in the Australian Pollen Record.

The amazing dominance of *Eucalyptus* over a large part of the Australian continent is immediately apparent. With about 450 species comprising forest giants, trees of all sizes, and shrubby Mallee — and occupying all sorts of habitats from the snowline on mountains to the shoreline with roots in saltwater, in deserts and in swamps and floodplains — Eucalypts have come to symbolise Australia. If a continent can be said to be characterised by an aroma, then that of burning Eucalypt twigs and leaves, whether in campfires or in wildfires, would surely be the distinctive "Australian" smell.

312 Gumnuts, hard and leathery capsules. (Densey Clyne)

313 Eucalypt leaves and Nothofagus, the Antarctic Beech, from Berwick Quarry, Victoria. Age 20 million years. (John Webb)

314 Gumnuts, probably Tristania, the Brush Box. (John Webb)

315 Eucalyptus torquata, a yellow flowered gum, with red capsules and pointed operculae on the buds. (Densey Clyne)

316 Bright red leathery capsule of a Mallee, Eucalyptus stoatei. (Densey Clyne)

317 Eucalyptus sieberi, a white flowered gum. (Densey Clyne)

312

313

314

315

316

317

318

318 *A sedge-like plant, this member of the family* RESTIONACEAE *is in flower with its white anthers hanging out of its bracts in the flowering spikes. (Peter Valder)*

319 RESTIONACEAE *and* PROTEACEAE *growing on sandstone in the Cape, South Africa, in a "fynbos" community. Both families have a Gondwanan distribution pattern. (Mary E. White)*

FAMILY RESTIONACEAE

117

The RESTIONACEAE, a family of sedge-like plants which grow on low nutrient soils and form tussocks or have creeping rhizomes is another of the old Gondwanan families. Its habitats range from sandplains under arid regimes to swamps with standing water for most of the year.

RESTIONACEAE is an important component of the Mediterranean-climate floras of southern Western Australia and the Cape Province of South Africa. All of its members are strongly adapted for drought resistance.

The different genera of RESTIONACEAE endemic to the southern continents clearly show that there has been evolution in isolation on parallel lines in the different lands, from a common ancestor which must have originated before the separation of Gondwana got underway (this implies a Late Cretaceous origin for the ancestor). It is particularly interesting that a family of herbaceous plants, grass-like, much modified for drought resistance or to adapt to poor soils, had such an ancient origin among the earliest Angiosperm families. There is a tendency to think of the first Angiosperms as being woody plants or even trees, and of the herbaceous habit as being a later development.

The radiation of RESTIONACEAE at the same time as the PROTEACEAE, FAGACEAE and other palaeoaustral families were being dispersed again emphasises that even when Rainforest was universal, niches were filled by plants already showing scleromorphic adaptation. (The Casuarinas are another example of sclerophyllous plants which appear very early in the Fossil Record.)

The adaptations in RESTIONACEAE made it well suited to colonise the sands of the Cretaceous epicontinental sea-beds when the waters had retreated and to populate the margins of water courses and swamps. (Grasses and sedges did not arrive on the scene in profusion to fill these niches until much later.)

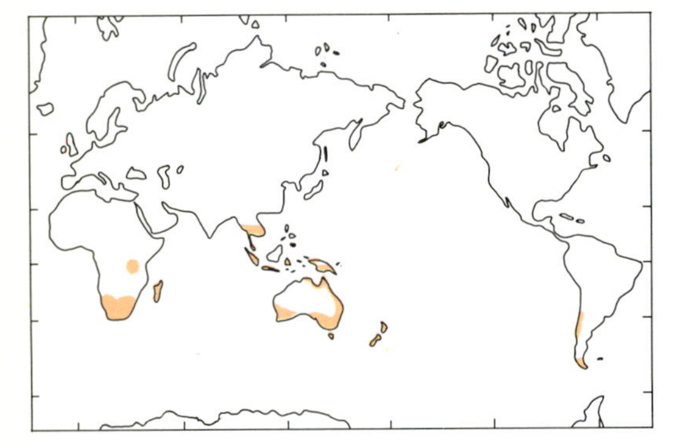

319

THE GRASSES

Grass pollen makes its appearance in the Pollen Record from the start of the Eocene, about 50 million years ago. It remains rare in eastern Australia until the Pliocene and only becomes abundant in the Pleistocene. However, in Central Australia it reflects the development of the arid region and is increasingly dominant from the Mid Miocene.

The tussock (*Astrebla*) and hummock (*Triodia* and *Plectrachne*) regions — the "spinifex plains" of the dry lands where soils are poor — are the sclerophyll-xerophyte expression of grassland. With the change from Closed Forest to open woodland as the climate became more seasonal and then increasingly dry, grasses became part of dry sclerophyll. The Arid Zone flora derived its components from dry sclerophyll. Every continent solved its desert problems in a different way: Africa with succulents and Thorn Trees; South America with Cactus and spiny shrubs; Australia with "spinifex", Saltbush and non-spiny *Acacias*.

In less arid regions than the Centre, the grasslands of tropical western Queensland and the Northern Territory are a major vegetation unit.

The production of open plains and grassland changed the previously short hopping jumps of some early Marsupials into the long, rhythmic hopping stride of the Kangaroos, creating the Antelope-equivalent in the Australian fauna. The "spinifex" plays a vital role in desert ecology, each clump providing a microclimate and a habitat for the desert dwellers — the Hopping Mice, Reptiles, Insects and other animals which live in the hot and dry Arid Zone.

320 A "spinifex" clump. (Glen Carruthers)

NEXT PAGE

321 Grassland near Armidale, New South Wales, in the late summer. (Jim Frazier)

322 South Coast, New South Wales. Rich grassland in a well watered district. (Densey Clyne)

323 Hummock grasses and Acacias in an arid-affected interzone area, Kinchega National Park. (Glen Carruthers)

324 Tropical North Queensland grassland with termite mounds. Paperbark forest behind, in an area which becomes swamp in the Wet. (Densey Clyne)

320

321
322

323

324

FAMILY SAPINDACEAE

Pollen of SAPINDACEAE appears in the record at the start of the Eocene. It represents the Closed Forest component of this family which flourished in the Miocene and then declined with contraction of the Rainforest. 111

The sclerophyll derivative of the family, with pollen referable to the modern genus *Dodonaea*, enters the record in the Mid Miocene as the Rainforest members decline.

FAMILY CHENOPODIACEAE — THE SALTBUSHES

Chenopod pollen types occur in the Pollen Record possibly as far back as the Oligocene. They are plants specially adapted for salt-tolerance, and their presence before aridity set in on the large semi-desert areas which they now occupy is again an indicator that local conditions had existed which created specialised niches for a long time before the main expression of a vegetation type.

The Saltbush country now feeds millions of sheep.

325 *Bassia paradoxa*, Family CHENOPODIACEAE, a saltbush. Clothed in downy hairs and with succulent, reduced leaves, it is adapted to drought and heat, as well as to the poor soils of its habitats. *(Densey Clyne)*

325

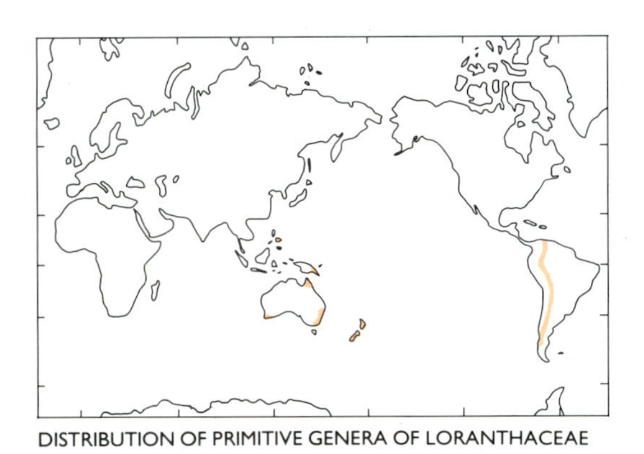

DISTRIBUTION OF PRIMITIVE GENERA OF LORANTHACEAE

326 *Nuytsia floribunda*, the Western Australian "Christmas Bush", is the largest-growing hemiparasite known. Its roots are parasitic on roots of other plants, but it makes its own food in the course of normal photosynthesis. *(Densey Clyne)*

327 Mistletoe flowers. Bunches of Mistletoe grow as parasites attached to branches of trees. They produce sticky seeds which are eaten by a bird (the Mistletoe Bird) whose activities are essential for the propagation of the parasite. The bird digests the flesh and expels the seed, ready to grow in its droppings which cling to the branch, usually gathering on the underside in a sheltered position. The germinating seed penetrates the bark with its rootlet, which enters the conducting tissue of the branch. The Mistletoe is tapped into its food supply and ready to grow, and may ultimately sap its host's strength. *(Densey Clyne)*

FAMILY LORANTHACEAE — THE MISTLETOES

The LORANTHACEAE is an old Gondwanan family. Its members are highly specialised in their modifications for their roles as hemi-parasites, in their embryology and in their methods of dispersal which involve Birds. They have low dispersibility and would have required land links to migrate over the parts of Gondwana.

A number of living, primitive genera show interesting distribution patterns which can only be explained by Gondwanan ancestry. In Australia there are *Nuytsia*, *Atkinsonia*, *Muellerina* and *Cecarria* — all endemic genera, monotypic or with a very small number of species. New Zealand has five primitive endemic genera and South America has six.

This pattern of restricted endemics which are relictual in nature is the classic expression of common ancestry in a progenitor which entered the different regions with the first radiations of Angiosperms. This radiation occurred prior to the separation of the continents and there has subsequently been parallel evolution in isolation.

The LORANTHACEAE probably evolved in Closed Forests under mild to warm conditions. Development of the hemi-parasite habit was their way of overcoming competition for light, food and water, short-circuiting the problem of poor soils.

Nuytsia, the Western Australian "Christmas Tree", is the largest hemi-parasite in the world. Its roots are parasitic on the roots of other trees.

326 327

FAMILY MIMOSACEAE, GENUS *ACACIA* — THE WATTLES

328

329

328 *Acacia triptera, a Wattle with prickly leaves. (Densey Clyne)*

329 *Acacia spectabilis, a Wattle with pinnate leaves. (Densey Clyne)*

330 *Most Acacias in Africa and India are "thorn trees", armoured with woody thorns. (Mary E. White)*

331 *Acacia longifolia, a Wattle with phyllodes. (Densey Clyne)*

Acacia is the biggest genus of trees and shrubs in Australia, with approximately 900 species. It and the Eucalypts are the most visible "Australian" plants and characterise the continent's vegetation.

Acacia pollen first appears in the record in Late Oligocene times, 25 million years ago. The plants are low pollen producers with limited dispersion and they are therefore not properly represented in fossil pollen samples. Their preference for drier habitats also mitigates against their presence in the Fossil Record.

The ancestors of modern *Acacia* must have been here for at least 25 million years before they appear in the Pollen Record. They were part of the Gondwanan stock present in Australia when it became isolated as an island-continent. The genus is widely distributed in Africa, South America and India. The onset of aridity and the development of sclerophyll prompted its great proliferation from the Miocene up to the Present.

Acacia is not a member of Rainforest communities. A few species occur in forests loosely classified as Rainforest, but the genus characterises open woodland and dry vegetation types. It largely replaces Eucalypts in the Arid Zone assemblages and occurs in Eucalypt-dominated wet sclerophyll and dry sclerophyll.

"Mulga Scrub" is the term used for the semi-desert vegetation characterising much of the continent's dry interior. Here Mulga (*Acacia aneura* and several other species) dominate, growing as low shrubby plants and forming a dense tangle with a few grasses and drought-tolerant herbs as ground cover, and scattered Eucalypts.

Many *Acacia* species have developed special adaptations to fire — lignotubers and fire-resistant seeds — as other members of the Sclerophyll have done.

The Australian Wattles are mainly of the section of the genus characterised by leaves which are "phyllodes" (a scleromorphic adaptation in which the petiole, or leaf stalk, is flattened into a leaf blade). Some species have the bi-pinnate compound leaves predominant in Africa's Thorn Trees. Others have juvenile leaves which are pinnate, followed by phyllodes in the mature state.

The Australian Acacias are almost all without thorns, in contrast to those of other lands. In Africa, India and South America nearly all the species of this genus are wickedly armoured with woody thorns.

330

MODERN AUSTRALIAN PLANTS WHOSE DISTRIBUTION SHOWS A GONDWANAN ORIGIN

Many modern Australian plants which do not have a fossil record are known to be of Gondwanan origin because of the distribution of their closest relatives in other southern lands. Those shown here are merely a sample of such plants, which reinforce the Gondwanan story.

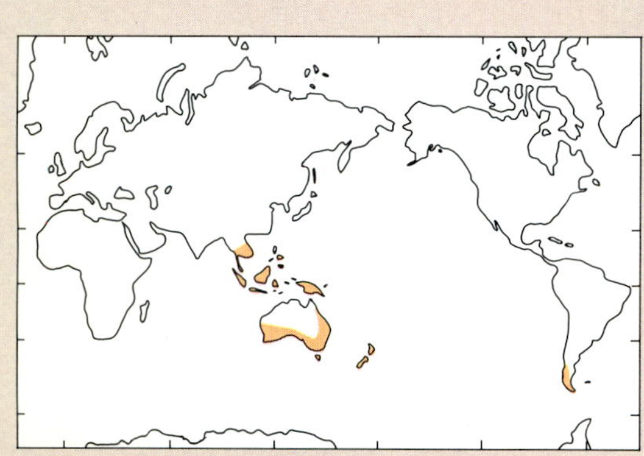

DISTRIBUTION OF EPACRIDACEAE

Family EPACRIDACEAE
This family is all Australasian except for an isolated occurrence at the tip of South America. It is particularly well developed in the Mediterranean flora of Western Australia. In the corresponding region in South Africa, the ERICACEAE fill this niche.

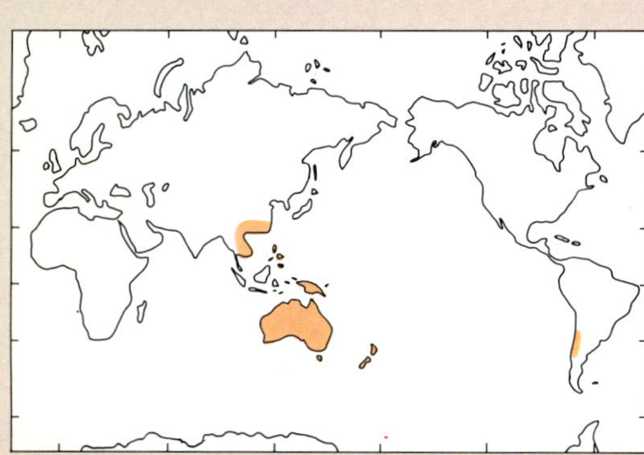

DISTRIBUTION OF GOODENIACEAE

Family GOODENIACEAE
This Australasian family has a limited area of distribution in western South America.

Family MYOPORACEAE
A small and homogeneous family, MYOPORACEAE is found mainly in Australasia, with outliers in the West Indies, South Africa, the Mascarenes and the extreme east of Asia.

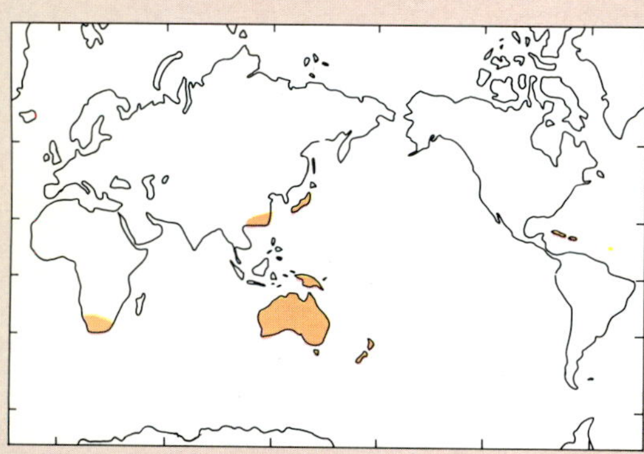

DISTRIBUTION OF MYOPORACEAE

Family RUTACEAE
This southern family has concentrations of genera in South Africa and Australia. One section of the family (the Citrus group of oranges, lemons, etc) extends into the Northern Hemisphere. Ancestral RUTACEAE were distributed in the early radiations of the Angiosperms, north and south from the region of origin. Development in isolation in Africa and Australia has resulted in parallel evolution of closely related endemic genera.

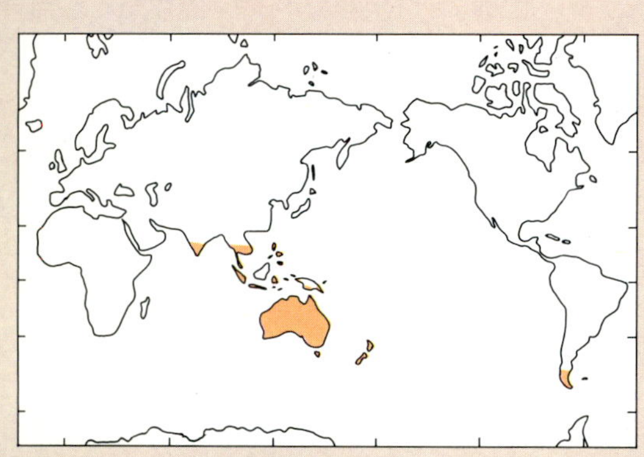

DISTRIBUTION OF STYLIDIACEAE

Family STYLIDIACEAE
Predominantly Australian, this family has a few species in eastern Asia, New Guinea, New Zealand, and sub-Antarctic South America. There is a concentration of many species in the Western Australian Mediterranean climatic zone. (See page 4)

332 *Epacris longiflora* (Native Fuschia), common on sandstone country in eastern Australia. Family EPACRIDACEAE. *(Densey Clyne)*

333 *Scaevola aemula*. Family GOODENIACEAE. *(Densey Clyne)*

334 *Eremophila willsii*. The generic name means "desert loving". Family MYOPORACEAE. *(Densey Clyne)*

335 *Boronia microphylla*. Family RUTACEAE. The foliage of Boronias contain aromatic oil glands. *(Densey Clyne)*

336 *Boronia crenulata*. *(Densey Clyne)*

332
333
334
335
336

AUSTRALIAN ENDEMICS

Endemic flora, which has evolved in isolation from Gondwanan stock, accounts for some 80 per cent of Australia's plant species and 30 per cent of the genera. Among the most unusual of these plants are the Grass-trees and the Kangaroo Paws.

Family XANTHORRHOEACEAE — the Grass-trees, Blackboys, Black-gins and Tinsel

This family is an endemic Australian, a highly evolved member of the sclerophyll community with nine genera. Of its 75 species, 70 are confined to Australia, and five occur in New Guinea and other close islands to our north (areas of land which are part of the Australian Plate or other areas that were in direct contact during times of lowered sea level and exposure of continental shelf during ice ages).

Genus *Anigozanthos* — the Kangaroo Paws

There is a very high proportion of endemism in the Mediterranean climatic region of Western Australia. *Anigozanthos manglesii* is the official State Flower.

337 A magnificent Grass-tree in a Western Australian landscape.

338 Flowering spikes of the "Black Boy". (Densey Clyne)

339 "Blackboys" in Dry Sclerophyll woodland. (Densey Clyne)

340 Anigozanthos, Kangaroo Paw. Occurring only in Western Australia.

337

338

339

340

* **Note:** *Generic and specific determinations of Tertiary Dicotyledon fossils illustrated follows Ettinghausen 1883. His choice of European taxa such as Fagus, Quercus, is outdated, and revision of Australian Tertiary plants is long overdue, the monographs of the last century being the only comprehensive works.* 120

341 Flower of *Brachychiton*.

342 A Bottle Tree *(Brachychiton)* near Goondiwindi, Queensland. *(Densey Clyne)*

343 Leaf of *Brachychiton sp.**, Family STERCULIACEAE, from Vegetable Creek, New South Wales. Age about 50 million years. (× 1.3)

341

342

343

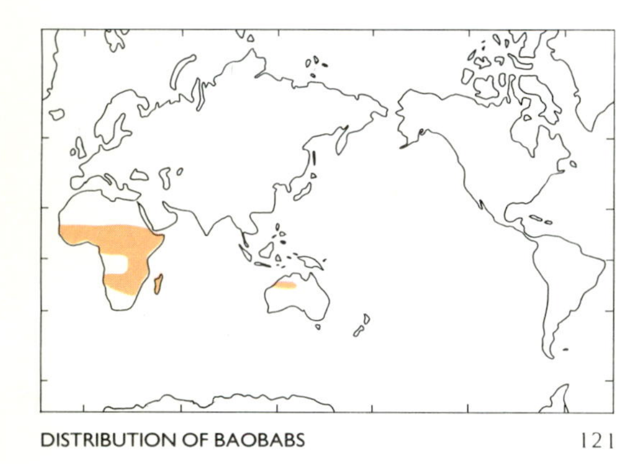

DISTRIBUTION OF BAOBABS 121

344

345

344 *"Bombax sturtii"** (left) and *Magnolia brownii** (larger leaf) from Dalton, New South Wales. (× 0.9)

345 *Baobab* trees.

346 *Lomatia brownii**, from Vegetable Creek, New South Wales. (× 1.9)

347 *Lomatia sp.**, from Dalton, New South Wales. Age 50 million years. (× 2.2)

348 *Lomatia silaifolia*. Regrowth after fire from underground lignotubers. *(Peter Valder)*

346

347

348

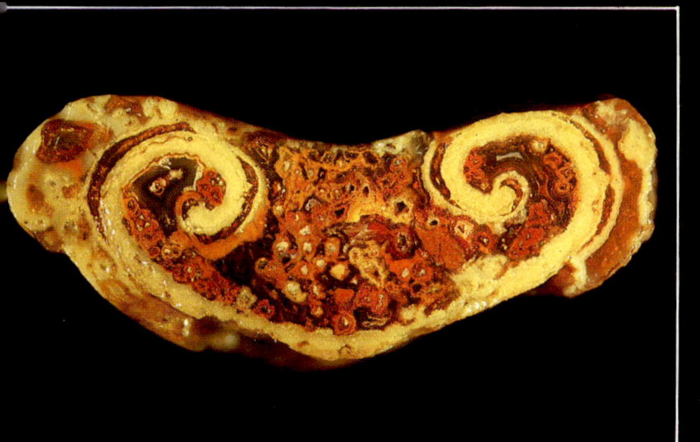

349

TASMANIAN PETRIFIED TREE-FERN TRUNKS

Polished sections of petrified Tree-fern trunk from the Lune River locality in Tasmania are famous for their beauty.

Although generally assumed to be of Tertiary age, 65 million years or younger, there is some doubt about this dating. The fossils occur below a basalt flow, which has not yet been dated accurately though geologists consider that it may be a Jurassic flow related to the Gondwana break-up. If, indeed, the basalt is this old, the Tree-fern fossils could be from the Jurassic or Triassic. It is necessary, therefore, to be cautious about their age until real evidence is forthcoming. Tree-ferns have been around for a very long time, since the Permian in Australia, and because of their long range they tell us nothing definite about their age.

The compound nature of Tree-fern trunks — consisting essentially of the vascular supply to the main axis, surrounded by a mass of vascular traces which supply the fronds, and interspersed with vascular traces to aerial roots, sometimes with the addition of an infilling of scales or fibrous tissue — results in spectacular petrifactions.

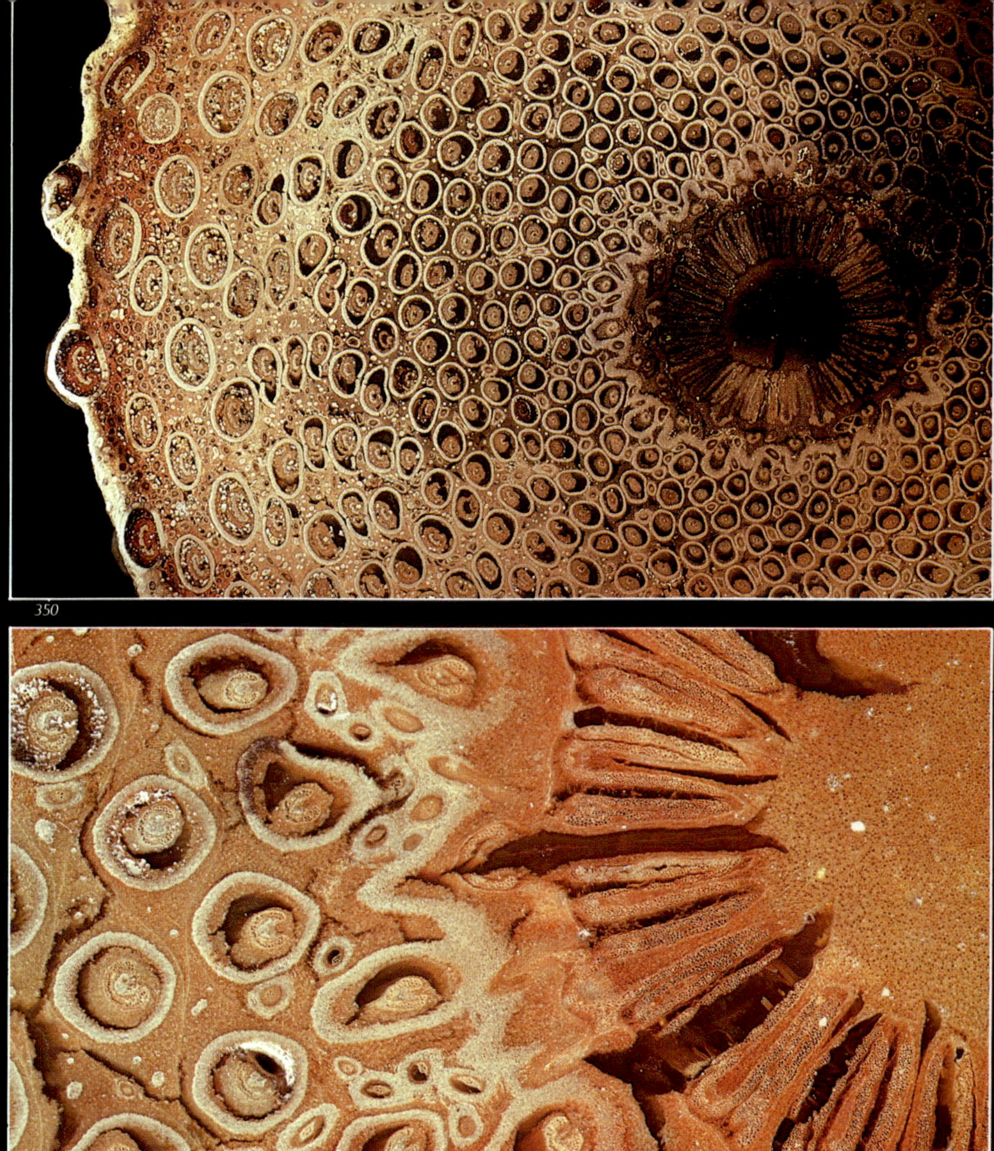

350

351

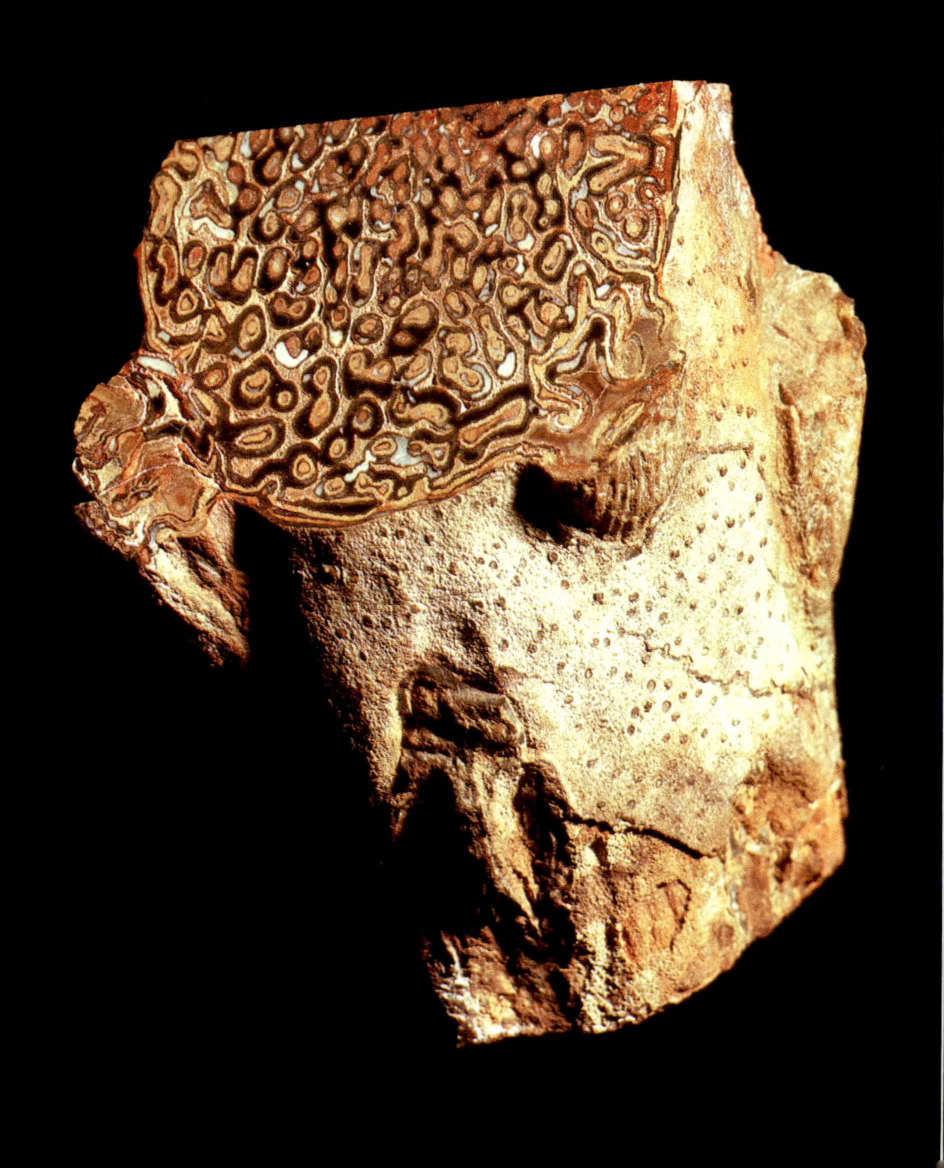

352

349 Rachis (leaf-stalk) of a Tree-fern frond, showing the elegant shape of its vascular trace.

350 Palaeosmunda sp. The vascular supply to the main axis is seen as a hollow ring of xylem at the right side of the picture. The spirally arranged leaf traces, getting larger towards the outside of the trunk, form an intricate pattern.

351 Palaeosmunda sp. Magnification of part of the vascular ring of tissue which supplied the main axis, and of vascular traces to fronds. The small spots between leaf trace bundles are roots in cross-section.

352 Angiopteris-like petrified Tree-fern. Rare, and as yet undescribed, petrified Tree-fern trunks in a collection owned by Mr Ron Smythe of Hobart, are of a different type from the osmundaceous examples common at Lune River, Tasmania. Their structure resembles that of living Angiopteris. This beautiful specimen is a completely preserved segment of stem showing the structure from outer layers inwards in its entirety. The stem is oval in cross-section, 4.5 x 3.5 cm. There is no core of vascular tissue to the main axis. Very numerous vascular bundles form a complicated pattern throughout. The outer layers of the stem form a smooth cover and the bases of the petioles form bosses. A pattern of small lumps all over the skin may be mucilage glands, a feature of living Angiopteris.

353 The complicated pattern of vascular bundles and leaf traces is here seen magnified.

354 Details of vascular traces to the petiole, which is a boss on the outside of the stem, are seen in this magnified view of part of the specimen.

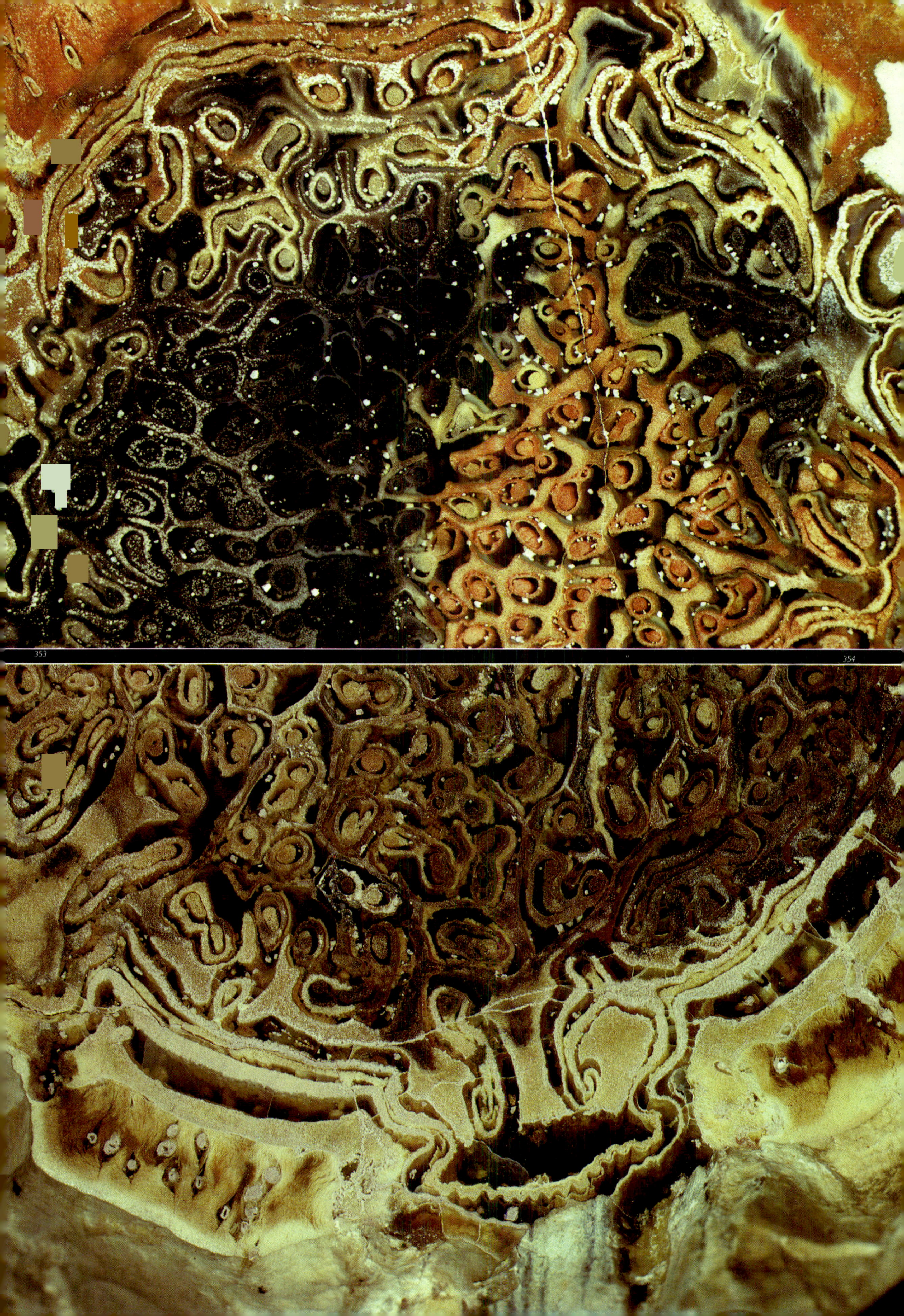
353

354

355 *Hakea bucculenta.*

356 *Hakea duttoni*, from Vegetable Creek, New South Wales.* (×3.0)

355

356

357

358

359

357 *"Fagus hookeri"**, from Vegetable Creek, New South Wales. (× 3.5)

358 *Roupala sapindifolia**, from Vegetable Creek, New South Wales. (× 2.5)

359 *"Piper feistmanteli"**, from Penrose, New South Wales. Age 50 million years. (× 0.8)

360 *"Quercus wilkinsoni"**, from Vegetable Creek, New South Wales. (× 1.6)

361 *"Quercus drymejioides"**, from Dalton, New South Wales. (× 1.2)

360

361

362 *Araucaria crassa*, a fossil Hoop Pine, from Ipswich, Queensland. Age, about 50 million years. (× 1.7)

363 *Phyllocladus asplenioides**, a fossil Celery-Top Pine, from Vegetable Creek, New South Wales. About 50 million years old. (× 1.9)

364 Dicotyledon leaf from Penrose, New South Wales. Age, about 50 million years. (× 1.8)

365 *Podocarpus cupressinoides**, from Penrose, New South Wales. Age 50 million years. (× 3.9)

366 *Cinnamomum sp.**, from Penrose, New South Wales. Family LAURACEAE. Age, 50 million years. (× 1.5)

367 *Cinnamomum leichardti**, Family LAURACEAE, from Elsmore, New South Wales. Age, about 50 million years. (× 1.9)

368 *Elaeocarpus muelleri**, Family ELAEOCARPACEAE, from Murrurundi, New South Wales. Age, about 50 million years. (× 3.1)

369 Leaf of a Dicotyledon, from Marawaka, New Guinea. Age, about 50 million years. (× 1.1)

362

363

365

367

366

368

364

369

CHAPTER 10

AUSTRALIA'S MODERN FLORA

THE QUATERNARY PERIOD

FROM 1.6 MILLION YEARS AGO UNTIL THE PRESENT

Climatic changes, more rapid and extreme than those of Tertiary times, had a profound effect on the vegetation. The flora was already well established and adapted, with plants suited to all niches. The climatic changes resulted in a shifting mosaic of communities similar to those of the present. Some admixture of northern plants with those derived from Gondwanan stock resulted from Australia's approximation to South-East Asia and had been assimilated into the Australian flora, particularly in the tropical north.

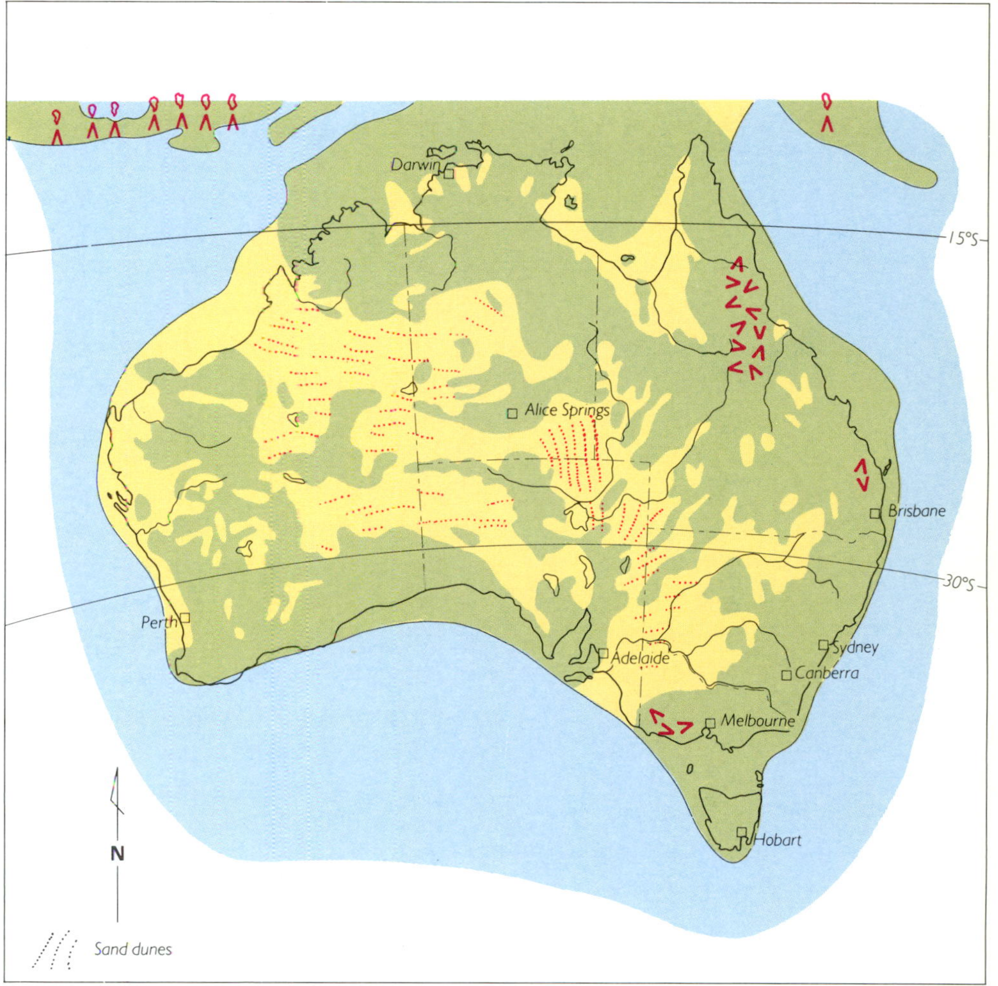

PALAEOGEOGRAPHY OF THE PLEISTOCENE.

The Closed Forests of the Early Tertiary contracted with the increase of seasonality of climate and the increasing aridity. The change from Closed Forests, to open woodlands, to grasslands and to deserts, with evolution of a sclerophyll-type vegetation that became increasingly arid- and fire-adapted, has been the main theme of the evolution of the modern Australian flora.

On the non-arid margins of the continent, as the areas suited to Rainforest contracted, the climate remained suitable for all grades of vegetation from wet sclerophyll, through dry sclerophyll, open woodland and grassland with all the local vegetation types we see today. Most of the Rainforest had been replaced by open vegetation dominated by Eucalypts and Wattles.

In the Quaternary, the climatic changes were generally more rapid and extreme than in the Tertiary and they affected the vegetation profoundly. As the flora was already well adapted and established, and the changes were within the range of tolerance of its plants, climatic changes did not mean extinctions and fundamental change but rather a shifting mosaic of communities similar to those of the present.

The waxing and waning of the polar ice sheets affected the landforms as well as the vegetation. During times of maximum extent, the ice caps were much larger than they are now. The cold episodes were arid, and the interglacial periods were warmer and wetter. The cycles of change can be seen in the Pollen Record, as vegetation composition changed with the climate. In Australia, small areas of ice and snow were confined to Tasmania and the Southern Alps.

When polar glaciation was at its peak the continental shelves were exposed and sea levels fell by up to 200 metres. Much of the Barrier Reef was a limestone plain similar to the Nullarbor today. Bass Strait became a lowland plain connecting Tasmania. In the north, plants and animals could migrate over almost continuous land.

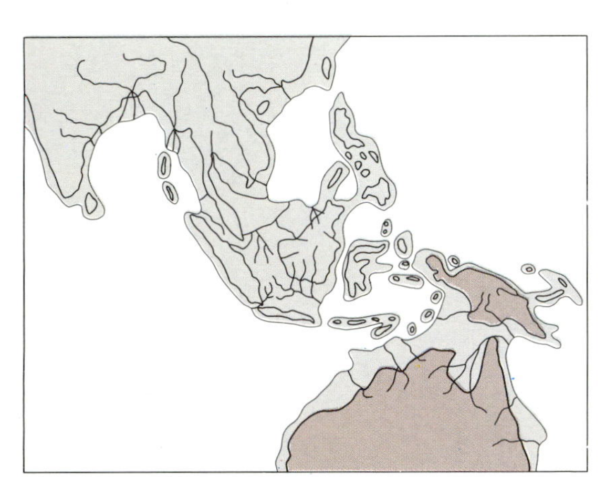

AUSTRALIA AND SOUTH EAST ASIA 120,000 YEARS AGO.
When polar glaciation was at its peak sea level fell as much as 200 metres and the continental shelves were exposed. Australia and New Guinea were almost united with South East Asia. 18

Cymbidium Orchid flowers, among the most advanced Monocotyledons.

MODERN AUSTRALIA

Australia is not only the smallest continent, it is also the flattest. Its long and stable 122 geological history has resulted in the widespread development of landscapes of plains and tablelands. Tectonically, it is a quiet continent, not subjected to volcanic activity or severe earthquakes. Its old rocks have been steadily eroded over the millennia, making it flatter as time went on.

Much of Western Australia consists of a plateau — the Old Plateau. Here the very old rocks have been exposed to almost continuous regimes of weathering.

The plains of Central Australia rise gently towards the east. The Great Divide, formerly called the Great Dividing Range, is not a mountain range, but a series of high plateaux ending abruptly in the Great Escarpment on its eastern margin. It stretches along eastern Australia from north of Cairns in Queensland to Victoria, with an extension across to Tasmania. Seen from the coast it looks like a mountain range because of the deep etching of the scarp by rivers and creeks. The uplift and creation of the steep seaward scarp are believed to have been associated with the opening of the Tasman-Coral seas during the breakup of Gondwana.

The high plains of eastern Victoria are areas of subdued relief at elevations of over 1500 metres, surrounded by terrain of much greater relief and deeper dissection. The land surface here originated at no great height above sea level in the Triassic Period and began to rise with the eastern highlands later in the Mesozoic.

The Great Escarpment separates the tablelands and the coastal belt. On the tablelands, relief is low and many of the old land forms are preserved because

TOPOGRAPHICAL MAP OF AUSTRALIA.

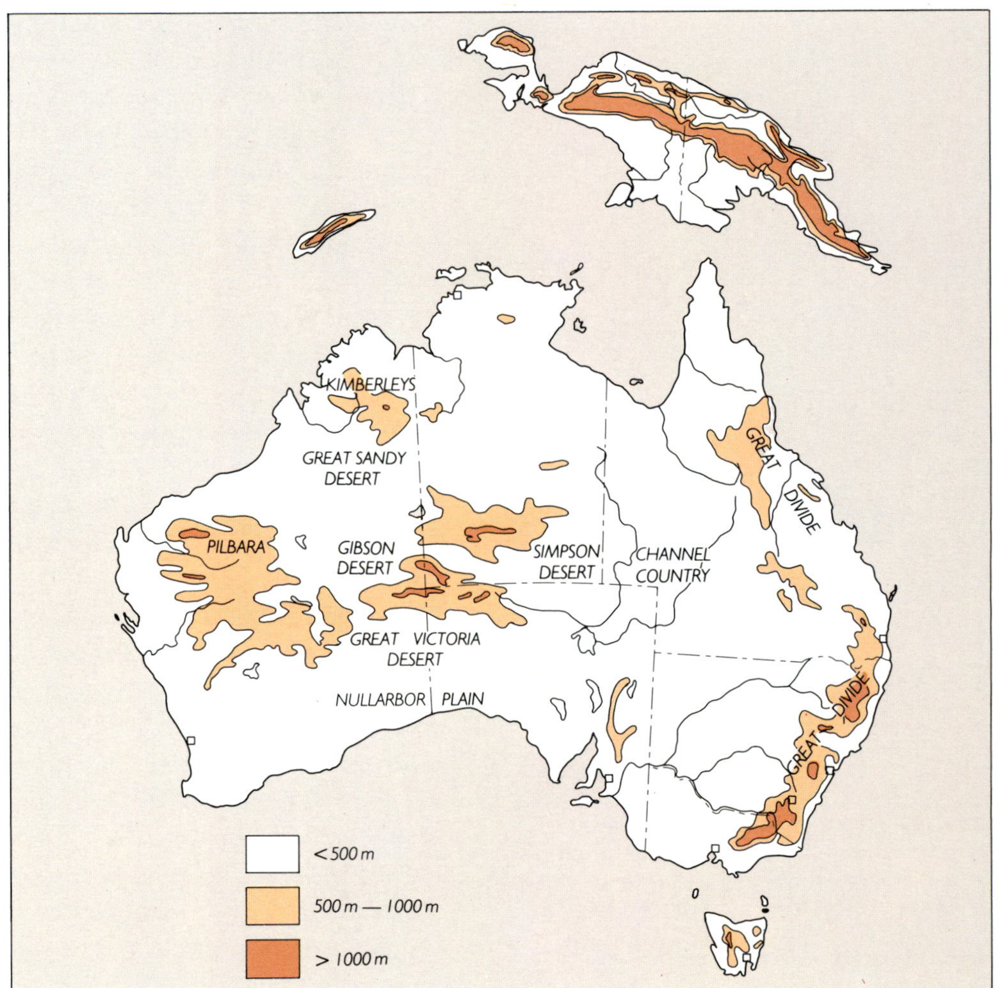

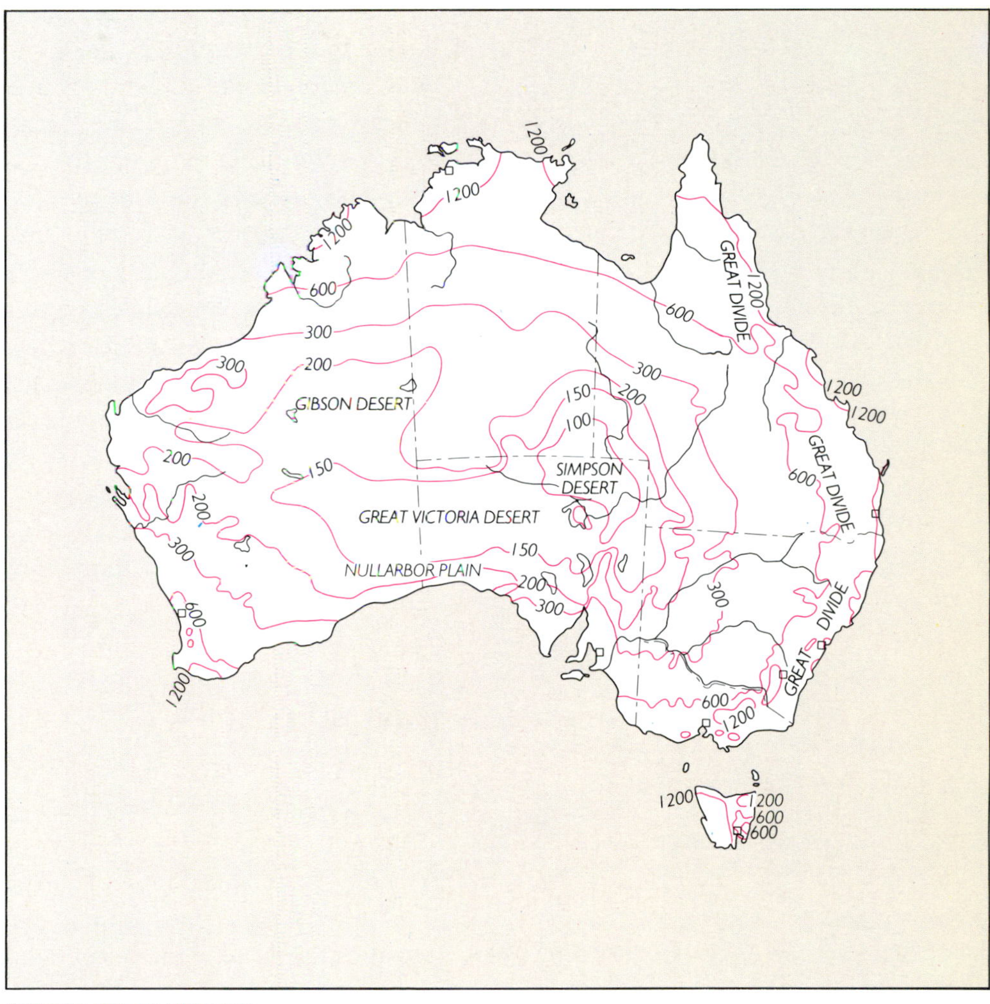

RAINFALL MAP OF AUSTRALIA.

erosion and tectonic movements have been slow and gentle. East of the scarp, relief is often great and geomorphic processes are rapid, thus old landforms are rare.

Slow erosion over vast periods of time, little tectonic activity and a landscape becoming even flatter as wide-spreading river systems broadcast sediments which fill in the shallow hollows on the surface, characterise the flat plains of the north and the Centre of the continent. The plain between Ayers Rocks and the Olgas is known to have existed much as it does now 60 million years ago, in Palaeocene times.

The proportion of arid regions in the continent had reached its present extent in Late Pliocene times, but true aridity with blown dunes only occurred in the Quaternary. About 20,000 years ago the interior became drier and windier. The desert conditions were even harsher than they are now, and the alluvium of the Early Pleistocene was blown into dunes.

Heavy metal deposits formed along the east and west coasts during the marine transgressions which followed the last ice age. Volcanic activity and the eruption of basalts occurred at isolated places in the Great Divide, notably in northern Queensland and in western Victoria. (The last eruptions were only a few thousand years ago.)

The present-day vegetation of the continent is varied, ranging from alpine regions to deserts and scrub.

ALPINE

Alpine habitats are of very limited extent in Australia. They are confined to altitudes above 1370 metres (4500 ft) in Tasmania and above 1830 metres (6000 ft) on the south-eastern part of the mainland. As in other parts of the world, they are characterised by the development of alpine moorland with *Sphagnum* bogs, sedges, heaths and Cushion Plants at highest altitudes. In the interzones between the moorland and the tree-line lower down, stunted and often gnarled and twisted trees and shrubs enter the communities.

On Mt Kosciusko there are ten alpine endemics including four kinds of Buttercup which have evolved in isolation in the relatively recent past, on their "islands" of high country. A number of "northern" genera are also found in the alpine floras and their presence is somewhat a mystery. They are probably migrants which travelled down the mountain chain from the islands to the north (moving down the Great Divide to their present situations in the south and facilitated in their spread, no doubt, by the cold periods during the ice ages).

ASTERACEAE, the Daisy family, is found in profusion and many of the sclerophyll genera have alpine species that exhibit the typical modifications of growth habit associated with such habitats (low growing, forming sprawling woody shrublets that cling close to the rocks for protection from wind and weather).

RAINFOREST

In general, the pockets of Rainforest found in Australia occur on suitable soils in topographically fire-proof niches, in sub-coastal regions with a mean annual rainfall of 800 millimetres in the tropical north and of 600 millimetres in the north-east. They represent "museums" of ancient plants.

In Early Tertiary times Rainforests clothed much of the continent. With the climatic changes and the onset of aridity that have occurred since then, they have contracted into refuges and European man has reduced the area of the Australian land surface which they formerly occupied from the 1 per cent which was the proportion on his arrival here to 0.25 per cent in 200 years. It is not difficult to see that Rainforest is highly endangered and will be extinct in the near future if steps are not taken to preserve what is left.

Rainforests are "Closed Forests", a name well chosen for they are closed ecosystems which function because their components, living and physical, are totally integrated. Their dynamics are destroyed when they are violated, be it by fire or by cutting roadways through them or by erosion on their margins allowing the stored nutrients in the soil to be dispersed and not recycled.

370 This rare petrified Tree-fern trunk from Lune River in Tasmania has a similar structure to Angiopteris. (See page 222 also)

ANGIOPTERIS

371 Angiopteris is a "living fossil". Today this rare King Fern is found in Madagascar, the Indo Malayan region, and in Australia, where it is found in Rainforest on Fraser Island, and in Carnarvon Gorge in Central Queensland. (Glen Carruthers)

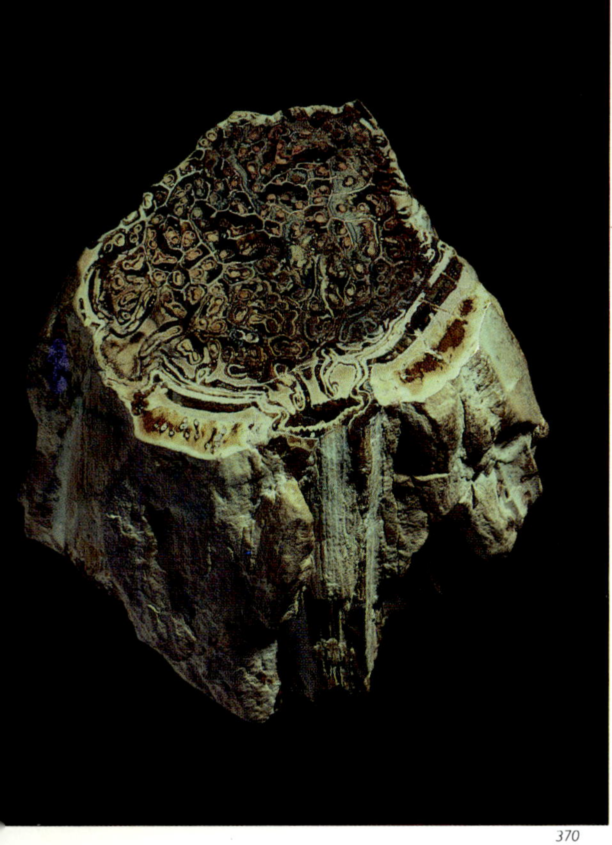

370

WET SCLEROPHYLL

Eucalypt species are dominant in the wet sclerophyll forests, whereas they are absent from true Rainforest. The Jarrah (*Eucalyptus marginata*) forests of Western Australia are prime examples of this forest type.

DRY SCLEROPHYLL

Dominated by Eucalypts, the dry sclerophyll areas are the "Australian Bush" as most of us visualise it — Gum-trees, Grass-trees, river banks with Paperbarks, Casuarinas and Tea-trees. This vegetation type is specially evolved to suit the low nutrient soils, to withstand droughts and to regenerate after fire.

372

373

372 Uluru (Ayers Rock) rises out of a flat plain. It represents extra hard and durable rock strata which remained when all the surrounding rocks were worn away and the plain created, and there is evidence that this landscape has changed little in 60 million years. Ephemerals, plants which appear and flower and die in a very short period after rain, carpet the ground, between the hummock grasses, scrub Acacias and Desert Oaks (Casuarinas). (Jim Frazier)

373 A flat country becoming even flatter. Erosion of a plateau gives rise to mesas — flat topped hills. (Densey Clyne)

374 A magnificent specimen of Ghost Gum (Eucalyptus papuana) in the Northern Territory. (Densey Clyne)

375 The "Red Centre" — old red soils, "spinifex" and Desert Oaks. (Densey Clyne)

WOODLANDS

The Australian woodlands are open Eucalypt areas and grassland, a mosaic of different communities determined by the local conditions of aspect, soil and topography. They grade into the increasingly arid vegetation zone of the interior of the continent, where wattles are co-dominants. Wattles become more frequent in the drier regions.

MALLEE WOODLAND AND MALLEE SCRUB

The Mallee are stunted Eucalypts growing as small trees or many-branched "whip stick" shrubs from lignotubers, adapted to drought and fire, with sparse grass herbs between the thickets. Features of the Mallee are its red, leathery fruit and its rapid regeneration after rain.

MULGA SCRUB

Stunted *Acacias* (Wattles) are dominant in the dry interior. They also occur with tussock and "spinifex" and some Eucalypts, Casuarinas and other sclerophylls in mixed-dominant situations.

SEMI-DESERT AND DESERT

Increasing aridity results in increasing desert adaptation. Mulga, Desert-oaks, "spinifex" and similar species are common, occurring in locally favourable pockets with other sclerophylls and xerophytes. Ephemerals carpet the ground after rain, their seeds lying dormant for long periods between times suitable for germination.

SALT-BUSH SCRUB

CHENOPODIACEAE is dominant in the scrub country, which is dry and surprisingly good for grazing, being able to feed hordes of sheep even on the margins of the deserts.

GRASSLANDS

In this vegetation type, trees are rare and grasses with an admixture of herbs and ephemerals dominate.

MANGROVES

Mangrove swamps along shorelines are of enormous ecological importance. They are breeding grounds for Fish and other marine and freshwater life. There is a tendency to disregard their function when they are situated close to populated areas, and to drain the swamps and reclaim the land. They must be preserved and managed with wisdom.

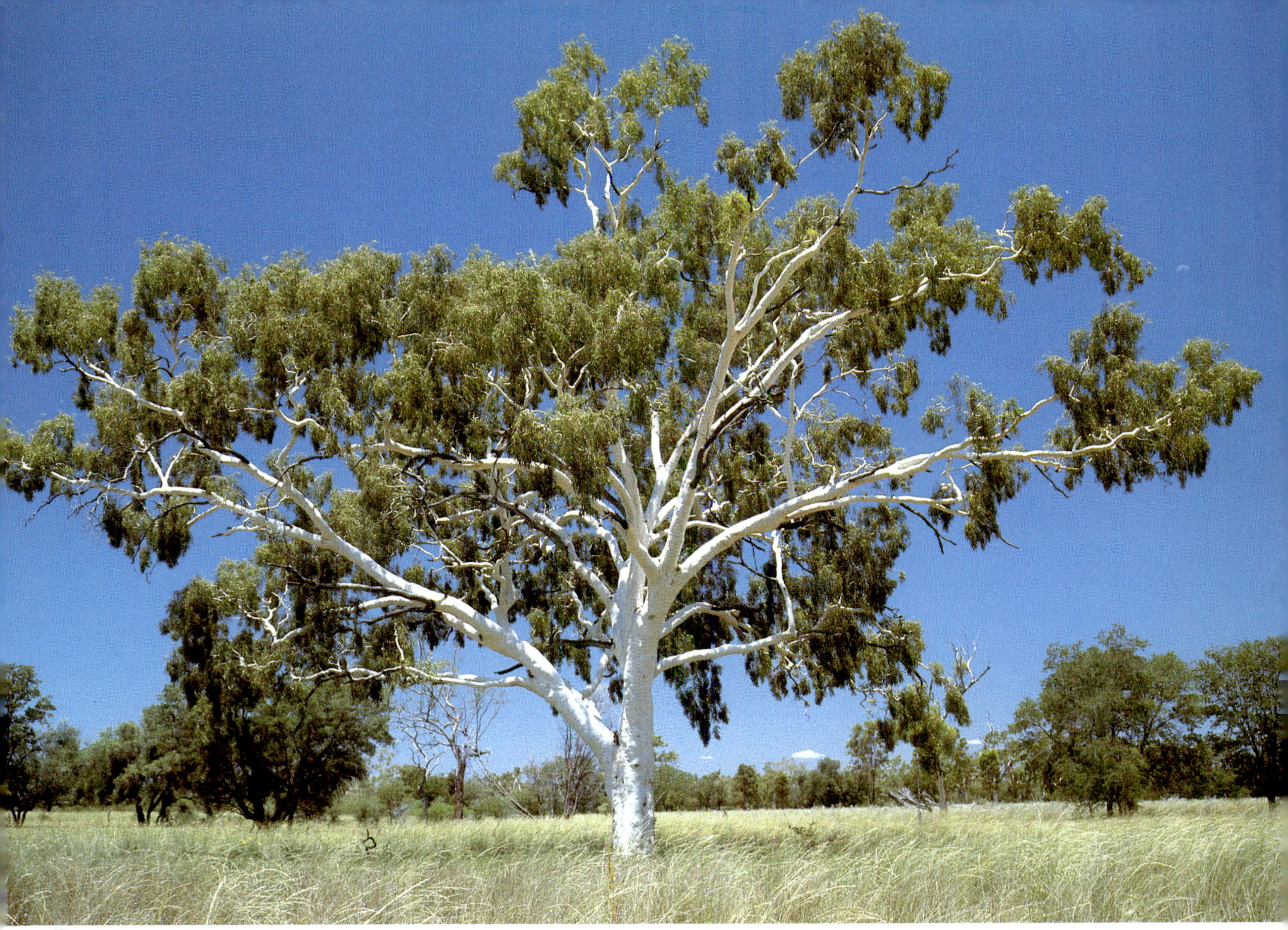

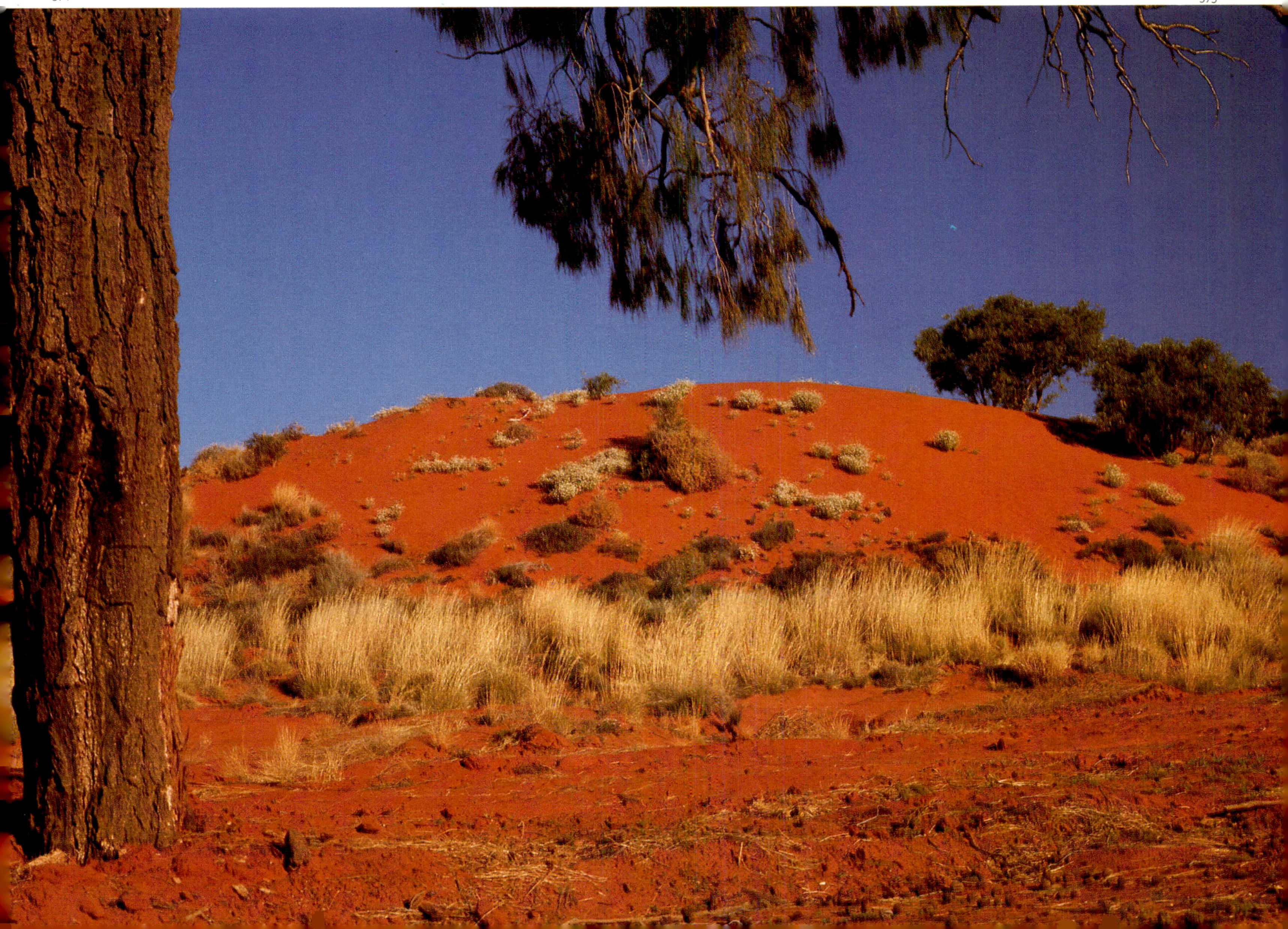

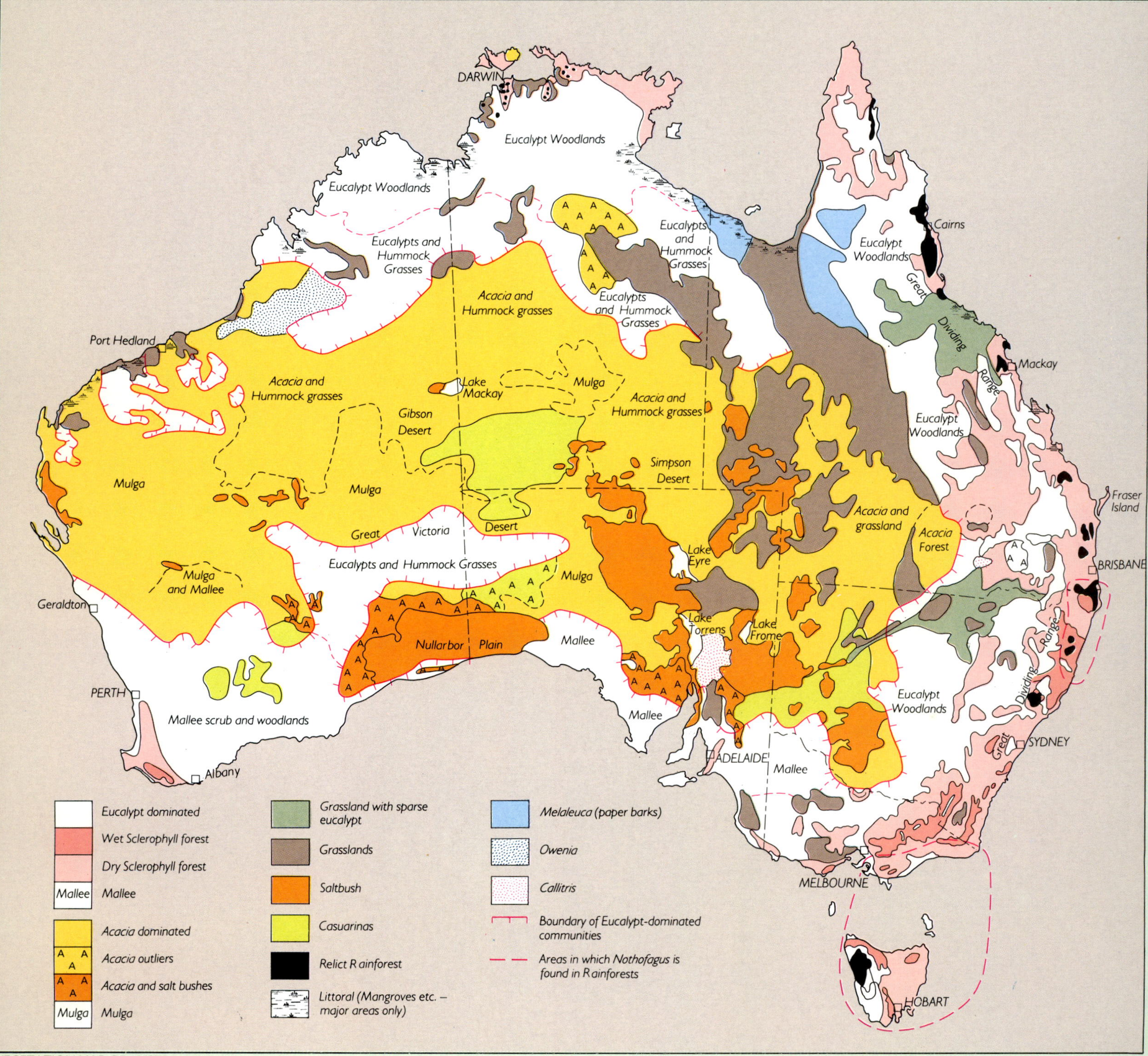
DARWIN
Eucalypt Woodlands
Eucalypt Woodlands
Eucalypts and Hummock Grasses
Eucalypts and Hummock Grasses
Eucalypts and Hummock Grasses
Cairns
Eucalypt Woodlands
Great Dividing Range
Mackay
Acacia and Hummock grasses
Port Hedland
Acacia and Hummock grasses
Lake Mackay
Mulga
Acacia and Hummock grasses
Gibson Desert
Eucalypt Woodlands
Simpson Desert
Mulga
Mulga
Fraser Island
Acacia and grassland
Acacia Forest
Great Victoria Desert
Lake Eyre
Eucalypts and Hummock Grasses
Mulga
BRISBANE
Mulga and Mallee
Geraldton
Lake Torrens
Lake Frome
Nullarbor Plain
Mallee
Great Dividing Range
PERTH
Eucalypt Woodlands
Mallee
Mallee scrub and woodlands
SYDNEY
ADELAIDE
Mallee
Albany
MELBOURNE
HOBART
Eucalypt dominated
Wet Sclerophyll forest
Dry Sclerophyll forest
Mallee
Mallee
Acacia dominated
Acacia outliers
Acacia and salt bushes
Mulga
Mulga
Grassland with sparse eucalypt
Grasslands
Saltbush
Casuarinas
Relict Rainforest
Littoral (Mangroves etc. – major areas only)
Melaleuca (paper barks)
Owenia
Callitris
Boundary of Eucalypt-dominated communities
Areas in which Nothofagus is found in Rainforests

VEGETATION MAP OF AUSTRALIA.

A NORTHERN INTRUSIVE ELEMENT IN AUSTRALIA'S FLORA AND FAUNA

As this book deals with Australia's Gondwanan ancestry, the invasion of the tropical north by plants and animals of Northern Hemisphere origin is beyond its scope. The complex interactions which started when the floating Australian Ark came into island-chain contact with South-East Asia (on the collision of the Australian Plate with Irian Jaya, and more significantly with the Timor region about 4 million years ago) is another story. The situation is enormously complicated.

It appears that far fewer "Australian" plants spread northwards than is the reverse case. A possible explanation for this situation is that the Australian flora is so finely adjusted to the soil-nutrient deficiencies which characterise most of its habitats that it doesn't adapt well and is unsuited to soils of regions to the north which are volcanic and comparatively rich.

The most significant "faunal" invasion from the north is undoubtedly Man. The Aboriginal Australians came here more than 30,000 years ago and became an integrated part of the continent's ecosystem, using its resources with intuitive wisdom and posing no long-term threat to its flora or its fauna. They were here when the giant Diprotodon grazed the land, when ice ages came and went, and when active volcanoes still erupted on the continent.

Because of Australia's isolation, surrounded by oceans and with only a scatter of islands chaining the continent to South-East Asia, it was "the ends of the Earth" when Britain was looking for somewhere to send unwanted citizens 200 years ago. It was disappointingly isolated and inhospitable for the earlier intrepid explorers of several nations searching for the "Great Southern Continent". They had hoped to find a landmass surrounding the South Pole, to "balance" the large continental masses of the Northern Hemisphere. Their *Terra Australis* was no land of milk and honey whose riches could be exploited by seamen-traders. They came and saw, and marvelled at the strange dry land with its weird assortment of plants and animals, and went away again — back to their familiar Northern Hemisphere with its softer greenness and its comforting predictability.

To the eyes of Captain Cook and his scientists and crew, accustomed to Britain's landscapes and vegetation, Australia must have seemed foreign and strange indeed. That the vegetation was recognised as being startlingly different from anything yet known is evidenced by the naming of "Botany Bay". There Joseph Banks made collections of the strangely different Australian plants, particularly *Banksias* which bear his name. Back in England the specimens and drawings created enormous scientific interest, which has continued unabated ever since, as has the interest in the equally weird and wonderful Antipodean fauna.

376

377

376 Banksia was named after Joseph Banks, the botanist who accompanied Captain Cook on his voyage in 1770. Banksia menziesi is a Western Australian species common around Perth and in the State's Mediterranean climate region, as well as on the coastal plains. (Densey Clyne)

377 Mature fruiting structure of a Banksia, charred black by a bushfire which caused its woody fruits to open and shed their seeds. (Densey Clyne)

THE FUTURE

When Modern Man came to Australia only 200 years ago, his impact was immediately felt. With him came his animals in their ever-increasing hordes, his foreign crops and weeds, his insatiable appetite for arable land and timber. And as he cleared the forests and altered the landscapes he knew little of what he was destroying, and thought little of the ultimate consequences of his actions.

What is left unspoilt is our heritage and our responsibility. How dare we destroy overnight the miraculous end-products of processes which were set in motion so many millions of years ago that we are unable to visualise time on such a scale. The theme throughout the ages has been co-operation and co-existence, the balance of give and take, the ability to adapt. Those that have been unable to adapt have become extinct.

Will Man, like Dinosaurs, be added to the list?

378 Isopogon latifolius, a member of the ancient Gondwanan family PROTEACEAE. *The back cover illustrate the specimen fully opened.*

GLOSSARY

Agate Semi-precious stone, a variegated form of chalcedony (which is a mixture of quartz and opal).

Algae (Latin *alga* means Seaweed.) The large group of Thallophyte plants containing chlorophyll, and thus able to synthesise carbohydrates. Blue-green, green, brown and red Algae, living in salt and freshwater.

"Algal Reef" (fossil.) Stromatolites. Limestones formed by activities of reef-forming Cyanobacteria (Cyanophytes).

Alternation of Generations Life cycle with a repeated pattern of asexual reproduction and sexual reproduction. (Diploid sporophyte plant produces haploid spores, which produce gametophyte plants which produce gametes. Gametes fuse to form diploid zygote from which the sporophyte develops.)

Angiosperms Vascular plants with seeds in ovaries — the Flowering Plants.

Annual Rings A circle in the stem of a plant marking a season's growth of wood.

Asexual Reproduction Reproduction without eggs and sperm.

Bacteria Common name for the Class Schizomycetes. Kingdom of microscopic organism.

Blue-greens Cyanophytes. Cyanobacteria.

Brown Algae Phaeophycophyta. Marine Algae characterised by a brown pigment.

Brown Coal An intermediate stage in the conversion of peat into true coal.

Bryophytes Mosses, Hornworts and Liverworts. Simple land-plants without vascular tissue or roots. These plants have a very long fossil history and were among the first plants to invade the land.

Bud An undeveloped shoot of a plant, often covered by scales.

Cainozoic The latest of the geological eras, from the end of the Cretaceous to the present day, thus including the Tertiary and Quaternary Periods. From 65 million years ago.

Cambium The tissue in roots and stems which is responsible for growth in diameter.

Cambrian The oldest group of rocks which contain recognisable fossils.

Carboniferous Geological Period between the Devonian and the Permian. From 360 to 286 million years ago. The age of Giant Clubmosses and Calamites in the Northern Hemisphere, which produced the coal deposits under tropical swamp conditions. No equivalent conditions or coal in the Southern Hemisphere during this Period.

Cast A fossil which is a mould of the outer form of a solid object like a root or a stem, without any preservation of the internal tissues. Internal casts can be produced in the case of hollow objects.

Chert A fairly pure siliceous rock.

Chlorophyll A green pigment essential for food manufacture in plants.

Chloroplast A specialised body within a cell, containing the chlorophyll.

Chromosomes Thread-like structures of cells which carry the genes.

Closed Forest A self-sustaining ecosystem which creates its own microclimate and recycles its own soil nutrients. Rainforests are Closed Forests.

Clubmosses The common name for Lycopods.

Coal A stratified rock consisting of the more or less altered remains of plants which have accumulated generally *in situ*. A fossil fuel.

Coal Measures Series of shales, sandstones, mudstones, etc, associated with coal seams. Result of sedimentation in a basin.

Cone A fruiting structure composed of modified leaves or branches which bear sporangia (microsporangia, megasporangia, pollen sacs, or ovules) and frequently arranged in a spiral or four-ranked order. For example, a Pine cone.

Conifer A cone-bearing plant.

Continental Drift The hypothesis that the present distribution of the continents is the result of fragmentation, followed by drifting apart of the fragments of a single concentrated landmass which is presumed to have existed in Carboniferous times.

Cordaites A group of early Gymnosperms assumed to be ancestral to most modern Conifers.

Cotyledon Seed leaf which stores food; two in Dicotyledons, one in Monocotyledons.

Cretaceous The geological Period between the Jurassic and the Tertiary. From 144 to 66.4 million years ago.

Cryptozoic Eon The Eon not characterised by life as we know it. Period of Earth's history up to 600 million years ago.

Cuticle Waxy layer on the outer wall of epidermal cells.

Cyanobacteria Bacteria-like photosynthesisers.

Cycadeoids Extinct group of plants related to modern Cycads.

Cycads Gymnosperms. From the Late Triassic to today. Woody plants, cone-bearing. Modern representatives are "living fossils" with many primitive characteristics, including a free-swimming sperm.

Cytoplasm The protoplasm of the cell, outside the nucleus.

Delta A roughly triangular area of river-borne sediment at the mouth of a river, heavily charged with detritus.

Deltaic Deposits Deposits of sand and clay with remains of brackish water organisms, drifted debris of plants and animals, washed from the land.

Desert An almost barren tract of land in which the rainfall is so scanty and spasmodic that it will not adequately support vegetation.

Detritus Fragmental material, such as sand and mud derived from the disintegration of rocks, and finely broken-up plant material.

Devonian Geological Period between the Silurian and the Carboniferous, from 408 to 360 million years ago.

Diamond The purest form of carbon known. The hardest mineral.

Dichotomous Branching by equal forking.

Dicotyledons Angiosperms whose embryos have two Cotyledons.

Diploid Having a double set of chromosomes, or referring to an individual containing a double set of chromosomes per cell. Usually a sporophyte generation.

DNA Deoxyribonucleic acid. The main constituent of chromosomes, carrying the genetic code.

Dropstones Rocks rafted on ice, often carried far from their source and deposited in sediments when the ice melts.

Dunes Ridges of sand which have been piled up by the action of wind on sea coasts or in deserts.

Ecology The study of plant and animal life in relation to environment.

Ecosystem A unit in Ecology. A specific habitat with its living organisms, which behaves as an integrated unit.

Egg A female gamete.

Embryo The structure formed after the egg has been fertilised. It grows into the new plant.

Embryophyta Major division of the Plant Kingdom.

Eocene Subdivision of the Tertiary Period. From 58 to 36 million years ago.

Erosion The lowering of the land surface by weathering, corrosion and transportation, under the influence of gravity, wind and running water.

Estuary The mouth of a river, where tidal effects are evident, and where fresh and sea water mix. An estuary usually develops in an area which has recently been submerged by the sea.

Eustatic movements Changes of sea level over wide areas, probably arising from variation in the quantity of water in the oceans due to the formation or melting of ice caps. The displacement of water by the accumulation of sediment may have the same effect, and seafloor spreading associated with movements of plates also causes changes in sea level.

Exine Outer coat of pollen grains.

Family Referring in plant classification to a group of genera.

Fault A fracture in the Earth's crust, along which the rocks on one side have been displaced relatively to those on the other.

Ferns Vascular plants which reproduce by spores. The sporophyte plants are diploid; their spores are haploid and produce gametophyte plants which bear male and female organs. Mobile sperm fuse with female gametes and a zygote is formed, and the sporophyte develops from it.

Fertilisation The stage of a sexual life cycle involving the fusion of male and female gametes, resulting in a diploid chromosome number.

Flower Floral leaves grouped together on a stem and adapted for sexual reproduction in the Angiosperms.

Fossil The remains and traces of animals and plants which are found naturally incorporated in rocks. They comprise not only the actual remains of organisms, such as bones, shells, leaves, stems, etc, but moulds and casts thereof, and impressions left by soft-bodied animals, leaves and plant fragments. Footprints of animals can be fossilised.

Frond A leaf of the sporophyte generation of a Fern.

Fruit A ripened ovary, with or without associated parts.

Fungi Thallophyte plants lacking chlorophyll and therefore deriving nourishment from other organisms.

Gametes Specialised reproductive cells that fuse in pairs to reproduce the next generation.

Gametophyte Plant generation that has sex organs and produces gametes.

Gene Particle contained in chromosomes that determines inherited characteristics.

Genus, plural Genera A group of living organisms, divided in turn into species.

Geological time The enormous span of time since the formation of Planet Earth. Divided into Eons, Eras and Periods.

Geology The science involving the deciphering of the whole evolution of the Earth and its inhabitants, from the time of the earliest records that can be recognised in the rocks to the present day.

Geomorphology The study of present-day landscapes, and the elucidation and explanation of their histories.

Giant Clubmosses The Lycopods of Late Devonian and Carboniferous times, which grew to great size.

Giant Horsetails Calamites, which like the Giant Clubmosses were very large.

Ginkgophytes Gymnosperms related to the modern Ginkgo or Maidenhair Tree. They have a long fossil history, and are now extinct except for one species.

Glaciation The covering of an area, or the action on that area, by an ice sheet or glaciers.

Glacier A mass of ice which moves slowly down a valley from above the snowline, under the force of gravity.

Glacigene Sediments Sediments produced by the action of glaciers.

Glossopterid, Glossopteridalean *Glossopteris* and its allies. An Order of Gymnosperms which characterise the Permian Period.

Gondwana The southern supercontinent which comprised the now separate continents of South America, Africa, Antarctica and Australia, as well as the sub-continent of India and the smaller landmasses of Madagascar, New Zealand, parts of Arabia, and probably fragments of south-eastern Asia.

Graptolites Extinct colonial animals, restricted to Ordovician and Silurian rocks, and of great value in stratigraphical correlations.

Green Algae Algae which have chlorophyll "a" and "b" in their cells. Ancestral to land-plants. (Seaweeds and fresh water.)

Gymnosperms Vascular plants with ovules not enclosed in ovaries, e.g. Conifers.

Habitat Specialised environment in which a specific organism lives.

Haploid Containing a single complete set of chromosomes, or referring to an individual or generation containing such a single set of chromosomes per cell. Usually the gametophyte generation.

Herb, Herbaceous Seed plant which does not develop woody tissues.

Heterosporous The condition of producing microspores and megaspores.

Homosporous The condition of producing one sort of spore only.

Horizon Level in a stratigraphical sequence.

Horsetails Plants with segmented stems, leaves in whorls at nodes, and cones which produce spores. Microscopic gametophyte generation. Important in the Fossil Record, now only one genus.

Hybrid Organisms produced by crossing two different species or varieties, or occasionally two genera.

Ice Age A period during which ice sheets and glaciers covered large areas of continents.

Igneous Rocks All rocks of magmatic and plutonic origin.

Impression Fossil Imprint of an organism on the surface of a layer of rock. No tissues of the organism are preserved.

Integument Protective outer layer of the ovule of seed plants.

Intertidal Zone Zone between the high and low tide marks on a shore.

Invertebrates Animals without backbones.

Jurassic Geological Period between the Triassic and the Cretaceous. From 208 to 144 million years ago.

Lacustrine Deposits Sediments which accumulated in a lake.

Lamina Blade, or expanded part of a leaf.

Laurasia A collective name for the area including North America and most of Asia and Europe. The northern supercontinent.

Lava Molten rock or magma which issues from a volcanic vent or fissure.

Leaching The process whereby material such as mineral salts and organic matter is washed out of a layer of rock and transferred to another by percolating water. Also applies to soils.

Lichen A Thallophyte composed of an Alga and a Fungus living together for their mutual advantage.

Lignite A variety of coal, intermediate in properties between the peats and the bituminous coals. Distinguished from brown coals by their more obvious organic matter.

Lignotuber A woody root stock from which the stems and branches of a plant grow. An adaptation to fire enabling a plant to regenerate from its underground organ when the aerial parts have been destroyed. Also a drought surviving strategy.

Lithosphere The solid crust of the Earth.

Liverworts A Class of the Bryophytes.

Ludlow The uppermost of the three divisions of Silurian time in Europe.

Lycopods Clubmosses. Reproduce by spores. Vascular plants with fossil history going back to the first land-plants 420 million years ago. Giant Clubmosses important in the Fossil Record. Now only a few genera of moss-like plants are alive, representing this important ancient Division of plants.

Macrofossil Fossil visible to the naked eye.

Magma Molten fluids, highly charged with gases and vapours generated in the depths of the Earth. Igneous rocks are formed from the consolidation of magma.

Mallee Shrubby Eucalypts with numerous stems, often forming thickets. Fleshy leaves and fruit, frequently brilliant red.

Mammals Hairy vertebrates, whose young are nourished on milk.

Marsupials Mammals characterised by having a pouch in which to carry their young, born imperfect.

Megaspore Spore which gives rise to a female gametophyte.

Meiosis Two special cell divisions occurring once in the life cycle of every sexually reproducing plant and animal, halving the chromosome number and effecting a segregation of genetic determiners.

Mesa A flat-topped, table-like mountain which falls away steeply on at least three sides, and is formed from a plateau in arid regions.

Mesozoic The Era of geological time during which the rocks of the Triassic, Jurassic and Cretaceous systems were deposited. From 245 to 66 million years ago.

Microspore A spore which gives rise to a male gametophyte.

Mineral A naturally-occurring inorganic substance which possesses a definite chemical composition and definite chemical and physical properties.

Miocene A division of the Tertiary, from 23 to 5 million years ago.

Mitochondria Small, cytoplasmic particles, associated with intracellular respiration.

Mitosis Nuclear division, involving the appearance of chromosomes, their longitudinal duplication, and equal distribution of newly formed parts to daughter nuclei.

Monocotyledon Angiosperm whose embryo has one cotyledon.

Mosses Bryophytes. Have a very long fossil history, from the beginning of life on the land. Non-vascular plants, which are the gametophyte generation. The sporophyte generation is represented by the capsules which bear the spores.

Mudstone An unlaminated, indurated sedimentary rock, consisting of clay minerals and other constituents of the mud grade.

Mulga Shrubby *Acacia*, often growing from lignotuber. Many thin stems forming thickets.

Mycorrhizal Fungi Fungi in symbiotic relationships with roots of plants.

Nectary A nectar-secreting gland.

Net Venation Veins of a leaf blade visible to the naked eye, branching frequently and joining again, forming a network.

Node Slightly enlarged portion of a stem where leaves and buds arise, and where branches originate.

Nucellus Tissue inside the outer integument of the ovules of Gymnosperms and Angiosperms, corresponding to the sporangium in other plants.

Nucleus A division of the protoplast of a cell. Essential in cellular synthetic and developmental activities.

Oil Shale A fine-grained dark brown or black shale containing kerogen.

Oligocene A division of the Tertiary. From 36.6 to 23.7 million years ago.

Opal A form of amorphous silica. "Wood opal" is wood in which the cavities have been filled and replaced by opal.

Ordovician Period Geological period from 505 to 436 million years ago.

Orogeny Changes in the levels of the Earth's crust which involve folding, faulting and thrusting and result in mountain formation.

Ovary Enlarged basal portion of the pistil which becomes the fruit.

Ovulate Referring to a cone, scale or other structure bearing ovules.

Ovule A rudimentary seed containing (before fertilisation) the female gametophyte, with egg cell, all being surrounded by the nucellus and one or two integuments.

Pangaea The single landmass comprising Laurasia and Gondwana before separation into a northern and a southern supercontinent.

Palaeobotanist Botanist who studies fossil plants.

Palaeocene Division of the Tertiary. From 66.4 to 57.8 million years ago.

Palaeogeography Study of ancient, prehistoric landforms.

Palaeontologist Scientist who studies fossils. Generally restricted to animal fossils.

Palynology The study of fossil pollen and spores.

Peat An accumulation of plant debris which has remained incompletely decomposed. The first stage in the transformation of vegetable remains into the series brown coal, lignite, bituminous and anthracitic coal, and anthracite.

Permian Period Period between the Carboniferous and the Triassic. From 286 to 245 million years ago.

Petal One of the flower parts, usually conspicuously coloured.

Petiole Stalk of a leaf.

Petrifaction Type of fossil in which the animal or plant is turned to stone by molecular replacement of the tissues by minerals.

Petrified Forest Petrified tree trunks in growth position with roots. Fossilised *in situ* when buried by mud, ash, etc.

Phanerozoic Eon Eon characterised by the presence of living organisms.

Phloem Food conducting tissue, consisting of sieve tubes, companion cells, phloem parenchyma and fibres.

Photosynthesis A process in which carbon dioxide and water are brought together chemically to form a carbohydrate, the energy for the process being supplied by the sun. Oxygen is released as a by-product, and chlorophyll is involved in the process, so plants which lack chlorophyll cannot manufacture their own food.

Pleistocene A division of the Quaternary, from 1.6 million years ago up to the last 10,000 years. It was the time of the great Ice Age, and there were four periods of intense cold alternating with four when conditions ameliorated and the glaciers and ice sheet retreated.

Pliocene The last division of the Tertiary, from 5.3 to 1.6 million years ago.

Pollen The male reproductive tissue of Flowering Plants.

Pollination The transfer of pollen from a stamen or staminate cone to a stigma or ovulate cone.

Prochlorons Primitive single cell Algae which have chlorophyll "a" and "b", and are believed to be ancestral to Green Algae, and hence to all Higher plants.

Proterozoic That part of the geological time column before life as we know it appears in the Fossil Record.

Prothallus The gametophyte plant of Ferns and other spore-bearing plants and the equivalent stage in Gymnosperms.

Protoplasm Living substance.

Psilophyta A Division of primitive early land-plants.

Pteridophytes Members of one of the main Divisions of the Plant Kingdom. Vascular plants with alternation of generations Ferns, Horsetails, Lycopods.

Pteridosperms Seed-ferns.

Pterophyta A Division of the Plant Kingdom comprising Ferns, Gymnosperms and Angiosperms.

Quartzite A granulose metamorphic rock, representing a recrystallised sandstone, consisting predominantly of quartz.

Quaternary The last 1.6 million years of the geological time column.

Rachis Main axis of a Fern frond.

Reproduction The process by which plants and animals give rise to offspring.

Rhizoid One of the cellular filaments which perform the function of roots.

Rhizome An elongated, underground, horizontal stem.

Rhizophore Structures in fossil Clubmosses which bear rootlets.

Rhyniophytes Early land-plants. Vascular, with terminal sporangia, ancestral to all higher plants except the Clubmosses.

Root The descending axis of a plant, normally below ground, and serving to anchor the plant and to absorb and conduct water and mineral nutrients.

Saprophyte An organism deriving its food from the dead body or the non-living products of another plant or animal.

Sedimentary Rocks Rocks formed of sediment, usually deposited in water.

Seed Formed from a fertilised ovule. Contains the embryo.

Sessile Lacking a stalk, as in a leaf without a petiole, or a flower or fruit without a pedicel, growing directly on a stem.

Sexual Reproduction Reproduction which requires meiosis and fertilisation for a complete life cycle.

Shale A laminated sedimentary rock composed of fine-grained particles.

Silurian The geological Period before the Devonian, from 436 to 408 million years ago.

Sporangium Spore case.

Spore A reproductive cell which develops into a plant without union with other cells — an asexual reproductive cell, as in Ferns.

Sporophyll A spore-bearing leaf.

Sporophyte The diploid plant, in alternation of generations, which produces spores by meiosis, halving their chromosome number.

Stomata Pores regulating the passage of air and water vapour to and from the inside of a leaf.

Strata Beds or layers of a sedimentary rock.

Symbiosis An association of two different kinds of living organisms resulting in mutual benefit.

Tertiary Geological Period from 66.4 to 1.6 million years ago.

Thallophytes A division of plants whose body is a thallus (lacking roots, stems and leaves).

Tracheid An elongated, tapering xylem cell, with lignified, pitted walls, adapted for conduction and support.

Tracheophytes Vascular plants.

Triassic Geological Period between the Permian and the Jurassic. From 245 to 208 million years ago.

Vascular Plants Plants containing vascular tissue, the conducting system which enables water and minerals to pass throughout the plant.

Venation The arrangement of veins through the leaf blade.

Vertebrates Animals with backbones.

Weathering The alteration and decay of rocks as far down as the depth to which atmospheric agencies can penetrate.

Xeromorphic Drought adapted.

Xerophylly Adaptation of leaves to dry conditions by reduction of lamina, thick cuticle, etc.

Xylem The woody tissues of roots and stems which conduct water and dissolved minerals upward.

Zone A group of strata characterised by a definite assemblage of fossils.

Zosterophylls Early, primitive land-plants, ancestral to the Clubmosses. Characterised by lateral sporangia.

Zygote Cell produced when two gametes fuse. The first cell of the new embryo plant. (The fertilised egg cell.)

BIBLIOGRAPHY AND FURTHER READING

1. Nieuwland, D. A. & Compston, W. 1981 Crystal Evolution in the Yilgarn Block near Perth, Western Australia. Evidence of age from detrital zircons: 2nd Intern. Symp. Archaean Geology, Perth, 1980. *Geol. Soc. Aust. Publ.* 7, p. 159-171.
2. Walter, M.R., Buick, R. & Dunlop, J. S. R. 1980 Stromatolites 3400-3500 Myr. old from the North Pole area, Western Australia. *Nature* 284; 443-445
3. Orpen, J. L. & Wilson, J. F. 1981 Stromatolites at ± 3500 Myr. and a Greenstone-Granite unconformity in the Zimbabwean Archean. *Nature* 291; 5812; 218-220.
4. Robertson, W. A. 1960 Stromatolites from the Paradise Creek area, N.W. Queensland. *Bur. Miner. Res. Aust. Report* 47.
5. Playford, P. E. a 1980 Environmental controls on the morphology of modern stromatolites at Hamelin Pool, W.A. W.A. *geol. Surv. Ann. Rep.* (for 1979); 73-79.
   b 1980 Australia's stromatolite stronghold. *Nat. Hist. Mag.* 89; 58-61.
6. Logan, B. W. 1961 *Cryptozoon* and associated stromatolites from the Recent, Shark Bay, Western Australia. *J. Geol.* 69; 517-533.
7. Larkum, A. W. D. & Barrett, J. 1983 Light Harvesting Processes in Algae, in *Advances in Botanical Research*, Woolhouse, H. W. ed. Academic Press.
8. Truswell, E. M. in press Antarctica — A history of terrestrial vegetation: *In* Iuyen, R. J. *Geology Antarctica*, Oxford Univ. Press.
9. du Toit, A. L. 1937 Our Wandering Continents. Oliver & Boyd. Edinburgh.
10. Wegener, A. 1924 The origin of continents and oceans. Methuen. London.
11. Webb, L. J. & Tracey, J. G. 1981 The Rainforests of Northern Australia, in *Australian Vegetation* (Groves, R. H. ed.) Cambr. Univ. Press.
    1981 Australian Rainforests. Patterns and Change, in *Ecological Biogeography of Australia* (Keast, A. ed.) Junk. The Hague.
12. Barlow, B. A. 1981 The Australian flora: its origin and evolution. Introduction to *Flora of Australia* 1, 25-75.
13. Beadle, N. C. W. 1966 Soil phosphate and its role in moulding segments of the Australian flora and vegetation, with special reference to xeromorphy and sclerophylly. *Ecology* 47; 992-1007.
14. Burbidge, N. T. 1960 The Phytogeography of the Australian region. *Aust. J. Bot.* 8, 2; 57-212.
15. Gill, A. M. 1975 Fire and the Australian Flora. A Revision *Aust. Forestry* 38; 4-25.
16. Bradstock, R. 1981 Our Phoenix flora. *Aust. Nat. Hist.* 20, 7; 223-226.
17. McNamara, K. J. & Scott, J. K. 1983 A new species of *Banksia* (Proteaceae) from the Eocene Merlinleigh Sandstone of the Kennedy Range, Western Australia. *Alcheringa* 7; 185-193.
18. Wilford, G. E. 1983 Phanerozoic Palaeogeography. Text for 3 Map Sheets, BMR Earth Science Atlas. Commonwealth of Australia.
19. Lamont, B. 1982 Mechanisms for enhancing nutrient uptake in plants, with particular reference to Mediterranean South Africa and Western Australia. *Bot. Rev.* 48, 3; 597-689.
20. Chaloner, W. G. & Macdonald, P. 1980 Plants invade the land. *Roy. Scot. Mus.* Edinburgh.
21. Banks, H. P. 1970 *Evolution and Plants of the Past.* Fundamentals of Botany Series. Macmillan.
22. Sporne, K. R. 1962 *The Morphology of Pteridophytes.* Hutchinson Univ. Library.
23. Bassett, M. G. & Edwards, D. 1982 Fossil Plants from Wales. *Nat. Mus. Wales*, geol. ser. 2.
24. Thomas, B. 1981 *The Evolution of Plants and Flowers.* Peter Lowe.
25. Banks, H. P. 1973 Palaeogeographic implications of some Silurian-Early Devonian Floras, in *Pap. 3rd Gondwana Symp. Canberra*, (Campbell, K.S.W. ed.).
26. Kidston, R. & Lang, W. H. 1917 Old Red Sandstone Plants showing structure, from the Rhynie Chert Bed, Aberdeenshire.
    1. *Rhynia*
    1920 2. *Asteroxylon*.
    Trans. Roy. Soc. Edinb. 51 and 52.
27. Lyon, A. G. 1964 The probable fertile region of *Asteroxylon mackeyi* Kidston & Lang. *Nature* 203; 1082, 1083.
28. Lang, W. H. & Cookson, I. C. 1935 On a flora, including vascular land plants, associated with *Monograptus* in rocks of Silurian age, from Victoria Australia. *Phil. Trans. Roy. Soc. Lond.* 221; 421-449.
29. Cookson, I. C. 1935 On plant remains from the Silurian of Victoria, Australia, that extend and connect floras hitherto described. *Phil. Trans. Roy. Soc. Lond.* 225; 127-148.
30. Lang, W. H. 1930 Some fossil plants of Early Devonian type from the Walhalla Series, Victoria, Australia. *Phil. Trans. Roy. Soc. Lond.* 219; 133-161.
31. Cookson, I. C. 1949 Yeringian (Lower Devonian) plant remains from Lilydale, Victoria, with notes on a collection from a new locality in the Siluro-Devonian sequence. *Mem. Nat. Mus. Vic.* 16; 117-131.
32. Lang, W. H. & Cookson, I. C. 1927 On some early Palaeozoic plants from Victoria, Australia. *Mem. Proc. Manch. Lit. Phil. Soc.* 71; 41-51.
33. Cookson, I. C. 1936 On occurrence of fossil plants at Warrentinna, Tasmania. *Pap. Proc. Roy. Soc. Tas.* 1936; 73-77.
34. Garratt, M. J. 1978 New evidence for a Silurian (Ludlow) age for the earliest *Baragwanathia* flora. *Alcheringa* 2; 217-224.
35. Garratt, M. J. et al. 1985 The appearance of *Baragwanathia* (Lycopsida) in a Silurian vascular flora. In press.
36. Tims, J. D & Chambers, T. C. 1984 Rhyniophytina and Trimerophytina from the early land flora of Victoria, Australia. *Palaeontology* 27; 265-279.
37. Walton, J. 1927 On some Australian fossil plants referable to the genus *Leptophloeum* Dawson. *Mem. Proc. Manch. Lit. Phil. Soc.* 70; 113-118.
38. Gould, R. E. 1975 The succession of Australian Pre-Tertiary megafossil floras. *Bot. Rev.* 41, 4; 453-483.
39. White, M. E. a 1973 Plant Fossils from the Drummond Basin, Queensland. *Bur. Miner. Res. Aust. Rep.* 11. 132; 73-78.
    b 1964 Plant fossils from the Emerald Sheet area. *Bur. Miner. Res. Aust. Rep.* 68; 68-71.
    c 1984 Australia's Prehistoric Plants. Methuen. Australia.
40. Walkom, A. B. 1928 Lepidodendroid remains from Yalwal, New South Wales. *Proc. Linn. Soc. NSW.* 53 (3); 310-314.
41. Dun, W. S. 1898 On the occurrence of Devonian plant-bearing beds on the Genoa River, County Auckland. *Rec. geol. Surv. NSW.* 85, 3; 117-121.

42. Morris, N. 1973 The *Rhacopteris* flora in New South Wales. *Pap. 3rd Gondwana Symp. Canberra*; 99-108.

43. Rigby, J. F. 1973 *Gondwanidium* and other similar genera and their stratigraphical significance. *Geol. Surv. Qld. Publ.* 350.

44. Archangelsky, S. & Arrondo G. 1971 Palaeophytologia Kurtziana 3, 2; Estudio sombre el genero *Botrychiopsis* Kurtz (= *Gondwanidium Gothan) del Carbonico y Permico Gondwanico. Ameghiniana* 8; 189-227.

45. Read, C. D. 1934 A flora of Pottsville age from the Mosquito Range, Colorado. *Prof. Pap.* U.S.G.S. 185 D.

46. Feistmantel, O. 1879-81 The Fossil Flora of the Gondwana System. *Mem. Geol. Surv. India: Pal. Indica.* Ser. 12, 3 (1-3).

47. Holmes, W. B. K. b 1981 A new leaf, *Glossopteris duocaudata* sp. nov., from the Late Permian of Cooyal, NSW. *Palaeobotanist* 28-29; 46-52.

48. Etheridge, R. a 1894 On the mode of attachment of the leaves or fronds to the caudex in *Glossopteris. Proc. Linn. Soc.* NSW. (ser. 2) 9, 2; 228-249.

49. Gould, R. E. & Delevoryas, T. 1977 The biology of *Glossopteris*: evidence from petrified seed-bearing and pollen-bearing organs. *Alcheringa* 1; 387-399.

50. Pant, D. D. 1977 The plant of *Glossopteris. J. Ind. Bot. Soc.* 56, 1; 1-21.

51. Plumstead, E. P. 1958 The habit of growth of Glossopteridae. *Trans. geol. Soc. Sth. Afr.* 61; 81-96.

52. Gould, R. E. 1973 A preliminary report on petrified axes of *Vertebraria* from the Permian of Eastern Australia, in *Gondwana Geology* (Campbell, K.S.W. ed) 3rd Gondwana Symposium. Canberra.

53. Feistmantel, O. 1890 Geological and palaeontological relations of the coal and plant-bearing beds of palaeozoic and mesozoic age in eastern Australia and Tasmania. *Mem. geol. Surv. NSW.* Pal. 3.

54. Plumstead, E. P. a 1952 Description of two new genera and six new species of fructification borne on *Glossopteris* leaves. *Trans. geol. Soc. Sth. Afr.* 55; 281-328.

b 1956 On *Ottokaria* the fructification of *Gangamopteris. Trans. geol. Soc. Sth. Afr.* 59; 211-236.

c 1956 Bisexual fructifications borne on *Glossopteris* leaves from South Africa. *Palaeontographica* B 100; 1-25.

d 1958 Further fructifications of the Glossopteridae, and a provisional classification based on them. *Trans. geol. Soc. Sth. Afr.* 61; 51-76.

e 1962 *Vannus gondwanensis* – a new *Gangamopteris* fructification from the Transvaal, South Africa. *Palaeobotanist* 11; 106-114.

55. Surange, K. R. & Chandra, S. a 1975 Morphology of the gymnospermous fructifications of the *Glossopteris* flora and their relationships. *Palaeontographica* B 149; 153-180.

b 1976 Morphology and affinities of *Glossopteris. Palaeobotanist* 25; 509-524.

56. White, M. E. a 1963 Reproductive structures in Australian Upper Permian Glossopteridae. *Proc. Linn. Soc. NSW.* 88, 3; 392-396.

57. Rigby, J. F. c 1978 Permian Glossopterid and other Cycadopsid fructifications from Queensland. *Geol. Surv. Qld. Publ.* 367.

58. Chandra, S. & Surange, K. R. 1977 Two new female fructifications – *Jambadostrobus & Venustostrobus* borne on *Glossopteris* leaves. *Palaeontographica* B, 164; 127-152.

59. Holmes, W. B. K. a 1974 On some fructifications of the Glossopteridales from the Upper Permian of NSW. *Proc. Linn. Soc. NSW.* 98; 131-141.

60. Walkom, A. B. a 1921 *Nummulospermum bowense* gen. et. sp. nov. *Quart. J. geol. Soc. Lond.* 77; 289-295.

b 1922 Palaeozoic floras of Queensland. Part 1. The flora of the Lower and Upper Bowen Series. *Geol. Surv. Qld. Publ.* 270.

61. Kovaks-Endrody, E. 1974 Seed-bearing *Glossopteris* leaves. *Palaeontologica Africana* 17; 11-14.

62. Walkom, A. B. 1935 Some fossil seeds from the upper Palaeozoic rocks of the Werrie Basin. *Proc. Linn. Soc. NSW.* 60, 5-6; 459-463.

63. Lacey, W. S. et al. 1975 Fossil plants from the Mooi River district of Natal, South Africa. *Ann. Natal Mus.* 22, 2; 349-420.

64. Surange, K. R. & Chandra, S. a 1973 *Partha* a new type of female fructification from the Lower Gondwana of India. *Palaeobotanist* 20; 356-360.

b 1973 *Denkania indica* gen. et sp. nov., a Glossopteridalean fructification from the Lower Gondwana of India. *Palaeobotanist* 20; 264-268.

c 1974 Some male fructifications of the Glossopteridales. *Palaeobotanist* 21, 2; 255-266.

d 1974 *Lidgettonia mucronata* sp. nov. a female fructification from the Lower Gondwana of India. *Palaeobotanist* 21, 2; 121-126.

e 1974 Further observations on *Glossotheca* Surange & Maheshwari. A male fructification of Glossopteridales. *Palaeobotanist* 21, 2; 248-254.

65. White, M. E. b 1978 Reproductive structures of the Glossopteridales in the Australian Museum. *Rec. Aus. Mus.* 31, 12; 473-505.

66. Thomas, H. H. 1958 *Lidgettonia*, a new type of fertile *Glossopteris. Bull. Brit. Mus. Nat. Hist.* 3; 179-189.

67. Pant, D. D. & Nautiyal, D. D. 1960 Some seeds and sporangia of *Glossopteris* flora from Raniganj Coalfield, India. *Palaeontographica* B 197; 41-64.

68. du Toit, A. L. 1932 Some fossil plants from the Karroo System of South Africa. *Ann. Sth.Afr. Mus.* 28 (4); 369-393.

69. Rayner, R. J. & Coventry, M. K. 1985 A *Glossopteris* flora from the Permian of South Africa. *Sth. Afr. J. Sci.* 81; 21-32.

70. Sahni, B. 1948 The Pentoxyleae, a new group of Jurassic gymnosperms from the Rajmahal Hills of India. *Mo. Bot. Gard. Ann.* 110; 47-79.

71. Meeuse, A. D. J. 1975 Aspects of the evolution of the Monocotyledons. *Acta. Bot. Neerl.* 24; 421-436.

1975 Floral evolution as the key to angiosperm descent. *Acta Bot. Indica.* 3; 1-18.

72. Burger, W. C. 1981 Heresy revived. The Monocotyledon theory of angiosperm origin. *Evolutionary Theory* 5; 189-225. Univ. of Chicago.

73. Doyle, J. A. 1978 Origin of the Angiosperms. *Ann. Rev. Ecol. Syst.* 9; 365-392.

74. Retallack, G. & Dilcher, D. L. 1981 Arguments for a Glossopterid ancestry of angiosperms. *Palaeobiology* 7, 1; 54-67.

75. Melville, R. 1983 Two new genera of Glossopteridae. *Bot. J. Linn. Soc. Lond.* 86; 275-277.

76. Rigby, J. F. 1967 On *Gangamopteris walkomii* sp. nov. *Rec. Aus. Mus* 27; 175-182.

77. White, M. E. c 1981 The cones of *Walkomiella australis* (Feist.) Florin. *Palaeobotanist* 28-29; 75-80.

78. Etheridge, R. 1899 On a fern (*Blechnoxylon talbragarense*) with secondary wood, forming a new genus, from the coal measures of the Talbragar District, NSW. *Rec. Aus. Mus.* 3; 135-147.

79. Gould, R. E. 1970 A new genus of siphonostelic Osmundaceous trunks from the Upper Permian of Queensland. *Palaeontology* 13, 1; 10-28.

80. Holmes, W. B. K. 1977 A pinnate leaf with reticulate venation from the Permian of New South Wales. *Proc. Linn. Soc.* NSW 102, 2; 52-57.

81. White, M. E. 1981 *Cylomeia undulata* (Burges), Gen. et. comb. nov. A lycopod from the Early Triassic strata of New South Wales. *Rec. Aus. Mus.* 33, 16; 723-734.

82. Retallack, G. 1975 The life and times of a Triassic Lycopod. *Alcheringa* 1; 3-29.

83. Burges, M. A. 1935 Additions to our knowledge of the flora of the Narrabeen Stage of the Hawkesbury Sandstone. *Proc. Linn. Soc.* NSW 60; 257-264.

84. Helby, R. J. & Martin, A. R. H. 1965 *Cylostrobus* gen. nov., cones of Lycopsidean plants from the Narrabeen Group (Triassic) of New South Wales. *Aust. J. Bot.* 13; 389-404.

85. Ash, S. R. 1979 *Skilliostrobus* gen. nov., a new Lycopsid cone from the Early Triassic of Australia. *Alcheringa* 3; 73-89.

86. Anderson, J. M. & Anderson, H. M. a 1983 Palaeoflora of Southern Africa. Molteno Formation (Triassic) 1. *Dicroidium*. Balkema Publ. Rotterdam.
b 1984 The fossil content of the Upper Triassic Molteno Formation, South Africa. *Palaeontologica Africana* 25; 39-59.

87. Townrow, J. A. 1957 On *Dicroidium*, probably a pteridospermous leaf, and other leaves now removed from this genus. *Trans. geol. Soc. Sth. Afr.* 60; 21-56.

88. Retallack, G. J. 1977 Reconstructing Triassic vegetation of Eastern Australia. A new approach for the biostratigraphy of Gondwana. *Alcheringa* 1; 247-277.

89. Holmes, W. B. K. 1982 The Middle Triassic flora from Benolong, near Dubbo, Central-western New South Wales. *Alcheringa* 6; 1-33.

90. Holmes, W. B. K. & Ash, S. R. 1979 An Early Triassic megafossil flora from the Lorne Basin, NSW. *Proc. Linn. Soc.* NSW. 103, 1; 48-69.

91. Thomas, H. H. 1933 On some pteridospermous plants from the Mesozoic rocks of South Africa. *Phil. Trans. Roy. Soc.* B. 222; 193-265.

92. Walkom, A. B. 1932 Fossil Plants from Mt Piddington and Clarence Siding. *Proc. Linn. Soc.* NSW 57, 3-4; 123-126.

93. Townrow, J. A. 1962 On *Pteruchus*, a microsporophyll of the Corystospermaceae. *Bull. Brit. Mus. Nat. Hist. Geol.* 6; 289-320.

94. Townrow, J. A. a 1956 The genus *Lepidopteris* and its Southern Hemisphere species. *Avhandl. Norske Vitensk. Akad.* Oslo.
b 1960 The Peltaspermaceae, a Pteridosperm family of Permian and Triassic age. *Palaeontology* 3, 3; 333-361.

95. Townrow, J. A. 1967 On *Rissikia* & *Mataia*, podocarpaceous conifers from the Lower Mesozoic of Southern lands. *Pap. Proc. Roy. Soc. Tas.* 101; 103-136.

96. Townrow, J. A. 1967 *Voltziopsis*, a Southern Conifer of Lower Triassic age. *Pap. Proc. Roy. Soc. Tas.* 101; 173-188.

97. White, M. E. 1981 Revision of the Talbragar Fish Bed Flora (Jurassic) of New South Wales. *Rec. Aus. Mus.* 33, 15; 695-721.

98. Walkom, A. B. 1921 Mesozoic Floras of NSW Part 1: Fossil plants from the Cockabutta Mt. and Talbragar. *Mem. geol. Surv.* NSW. Pal 12.

99. Walkom, A. B. 1941-42 Fossil plants from Gingin, WA. *J. Roy. Soc.* WA. 28; 201-207.

100. Douglas, J. G. 1969 Mesozoic Floras of Victoria. Parts 1 & 2.
1973 Mesozoic Floras of Victoria. Part 3. *Geol. Surv. Vic. Mem.* 28, 29.

101. White, M. E. 1961 Report on 1960 collections of Mesozoic plant fossils from the Northern Territory. *Bur. Miner. Res. Canberra*. Records 1961/146.

102. Walkom, A. B. 1919 Mesozoic floras of Queensland, 3 & 4. Floras of the Burrum and Styx River Series. *Qld. geol. Surv. Publ.* 263.

103. Doyle, J. A. 1977 Patterns of Evolution in early Angiosperms, in *Patterns of Evolution*. Hallam, A. ed. Elsevier Sci. Publ. Co. Amsterdam.

104. Burger, D. 1981 Observations on the earliest angiosperm development with special reference to Australia. *Proc IV int. palynol. Conf. Lucknow* (1976-77) 3, 418-428.

105. Martin, H. A. 1982 Changing Cenozoic Barriers and the Australian palaeobotanical record. *Ann. Missouri Bot. Gard.* 69; 625-667.

106. Stevens, G. R. 1980 *New Zealand Adrift*, Reed Publications, New Zealand.

107. Coetzee, J. A. & Praglowski, J. 1984 Pollen evidence for the occurrence of *Casuarina* & *Myrica* in the Tertiary of South Africa. *Grana* 23; 23-41.

108. Kemp, E. M. 1978 Tertiary Climatic evolution and vegetation history in the Southeast Indian Ocean region, in *Palaeogeography, Palaeoclimatology, Palaeoecology* 24; 169-208.

109. Martin, H. A. 1984 Australian Phytogeography, in *Vertebrate Zoogeography and Evolution in Australasia*. (Animals in space and time). Archer, M. & Clayton, G. eds. Hesperian Press.

110. Martin, H. A. 1981 The Tertiary Flora, in *Ecological Biogeography of Australia*. (Keast. A. ed.) Junk Publ. The Hague.

111. Martin, H. A. 1978 Evolution of the Australian flora and vegetation through the Tertiary. Evidence from pollen. *Alcheringa* 2; 181-202.

112. Martin, H. A. 1977 The history of *Ilex* (Aquifoliaceae) with special reference to Australia. *Aust. J. Bot.* 25; 655-673.

113. Truswell, E. M. 1984 The temperate Southern Beech/Podocarp Rainforest: Eocene to Miocene evolution in Australia, Antarctica and New Zealand. Pap. *6th Int. Palyn. Conf. Calgary*.

114. Morley, B. D. & Toelken, H. R. eds. 1983 Flowering Plants in Australia. Rigby, Publ.

115. Coetzee, J. A & Muller, J. 1985 The phytogeographical significance of some extinct Gondwana pollen types from the Tertiary of the South-west Cape, South Africa. *Ann Missouri Bot. Gard.* 71; 1088-1089.

116. Walker, J. W. et al. 1985 Winteraceous pollen in the Lower Cretaceous of Israel. *In* Origin and Early Evolution of Angiosperms. ed. C. B. Beck. *Science* 220.

117. Johnson, L. A. S. & Briggs, B. G. 1981 Three old Southern Families – Myrtaceae, Proteaceae and Restionaceae, in *Ecological Biogeography of Australia*. (Keast, A. ed) Junk Publ. The Hague.

118. Deacon, H. J. et al. eds. 1983 Fynbos palaeoecology; A preliminary synthesis. *Sth Afr. Nat. Sci. Programmes Report* 75.

119. Barlow, B. A. 1983 Biogeography of Loranthaceae & Viscacaceae, in *Biology of Mistletoes*. Academic Press.

120. Ettinghausen, O. 1883 *Tertiary Flora of Australia*.

121. Hutchinson, J. 1959 *The Families of Flowering Plants*. Clarendon Press, Oxford.

122. Ollier, C. D. 1977 Early Landform Evolution, in *Australia, A Geography* (Jeans, C. P. ed.) Univ. of Sydney Press.

123. Archer, M. & Clayton, G. 1984 *Vertebrate Zoogeography and Evolution in Australasia*. (Animals in space and time). Archer, M. & Clayton, G. eds. Hesperian Press.

TABLE OF ILLUSTRATED FOSSILS

Specimen Numbers: Prefixes indicate the Collections in which the fossils are housed:

F.: CPC:	Bureau of Mineral Resources, Canberra.
AMF:	The Australian Museum, Sydney.
MMF:	Mining Museum, Sydney.
MV:	Museum of Victoria, Melbourne.
BMNH:	British Museum of Natural History, London.
JD:	Collection of Dr J. Douglas, Melbourne.
LR:	Collection of Mr R. Smythe, Hobart, Tasmania.
MEW:	Collection of author.

Page	No.	Name and author of species	Fossil no.	Locality	Age
p. 10	5	*Dicroidium odontopteroides* (Morris) Gothan *var. moltenense* Retallack, and *Dicroidium elongatum* (Carruthers) Archangelesky *var. rigidum* (Dun) Stipanicic & Bonetti	AMF 61138	Benolong, NSW	Triassic
p. 22	16	North Pole stromatolite	AMF 63743	North Pole, W.A.	3500 m.y.
p. 26	19	*Dadoxylon farleyense* Walkom	AMF	Farley, NSW	Permian
p. 27	20	Opalised wood	AMF Galman Coll.	Lightning Ridge, NSW	Cretaceous
p. 27	21	Petrified tree trunk	AMF	New England, NSW	Tertiary
p. 28	22	*Palaeosmunda sp.*	L.R.	Lune River, Tas.	Tertiary
p. 29	23	*Palaeosmunda sp.*	L.R.	Lune River, Tas.	Tertiary
p. 30	24	*Agathis jurassica* White	MEW	Talbragar, NSW	Jurassic
p. 31	25	Opalised cones	AMF Galman Coll.	Lightning Ridge, NSW	Cretaceous
p. 31	26	*Leptophloeum australe* (McCoy) Walton	AMF 57155	Barraba, NSW	Late Devonian
p. 31	27	*Dicroidium* cuticle	MEW	South Africa	Triassic
p. 34	29, 30	*Araucarioxylon sp.*	MEW	Antarctica	Permian
p. 37	32	*Gangamopteris cyclopteroides* Feist.	AMF 58819	Antarctica	Permian
p. 38	33	*Glossopteris sp.*	AMF 58822	Antarctica	Permian
p. 38	34	*Dicroidium sp.*	AMF 4158	Argentina	Triassic
p. 39	35	*Dicroidium odontopteroides* (Morris) Gothan	AMF 32633	South Africa	Triassic
p. 38	36	*Glossopteris indica* Feist.	F 436	India	Permian
p. 38	37	*Dicroidium odontopteroides* (Morris) Gothan	AMF 58812	Horseshoe Mts., Antarctica	Triassic
p. 39	38	*Vertebraria indica* Royle	F 23266	Antarctica	Permian
p. 60	62	*Baragwanathia longifolia* Lang & Cookson	MV p 178123	Frenchmans Spur., Vic.	Late Silurian
p. 62	63	Psilophyte stems	CPC 4318	Einasleigh, N.E. Qld.	Devonian Broken R. Formation
p. 65	64	*Cooksonia sp.*	AMF 51793	Molong, NSW	Late Silurian
p. 66	66	Rhynie chert	MEW	Rhynie, Scotland	Early Devonian
p. 67	69	*Baragwanathia longifolia* Lang & Cookson	MV	Yarra Track, Vic.	Early Devonian
p. 68	70	*Baragwanathia longifolia* Lang & Cookson	MV p	Yarra Track, Vic.	Early Devonian
p. 68	71	*Baragwanathia longifolia* Lang & Cookson	MV p 178125	Yarra Track, Vic.	Early Devonian
p. 69	72	*Baragwanathia longifolia* Lang & Cookson	JD 259A	Yea, Vic.	Late Silurian
p. 70	73	*Hedeia sp.*	MV p 50033	Wilson's Creek, Vic.	Early Devonian
p. 70	74	*Pluricaulis biformis* Chambers & Tims	JD102a	Yea	Late Silurian
p. 70	75	*Yarravia sp.*	MV p 41819	Yarra Track, Vic.	Early Devonian
p. 71	76	*Dawsonites racemosa* (Lang) Hoeg	AMF 26066	Bowning, NSW	Late Silurian to Early Devonian
p. 71	77	*Palaeostigma sp.*	AMF 57153	Mudgee, NSW	Siluro-Devonian
p. 74	78	*Ulodendron* scars on *Lepidodendron sp.*	AMF 63755	Wallarobba, NSW	Carboniferous
p. 75	79	*Lepidostrobus sp.*	CPC 4344	Bowen Basin, Qld.	Carboniferous Base of Bulgonunna Volcanics
p. 75	80	*Lepidostrobus sp.*	CPC 4387	Bowen Basin, Qld.	Early Carboniferous Drummond Series

Page	No.	Name and author of species	Fossil no.	Locality	Age
p. 76	81	*Leptophloeum australe* (McCoy) Walton	BMNH	Broken R., N. Qld.	Late Devonian
p. 77	82	*Lepidodendron veltheimianum* Stbg.	F 22758	Drummond Basin, Qld.	Carboniferous Scartwater Fm.
p. 78	83	*Lepidodendron aculeatum* Stbg.	F 22946	Charters Towers, Qld.	Early Carboniferous
p. 78	84	*Lepidodendron mansfieldense* McCoy	F 22840	Drummond Basin, Qld.	Carboniferous
p. 78	85	*Lepidodendron sp.*	CPC 4388	Drummond Basin, Qld.	Early Carboniferous Scartwater Fm.
p.78	86	*Lepidodendron canobianum* Crookall	F75.30.0800	Georgetown Dist., Qld.	Carboniferous
p. 79	87	*Lepidodendron sp.*	F 22774	Drummond Basin, Qld.	Carboniferous Star of Hope Fm.
p. 79	88	*Stigmaria ficoides* Bgt.	F 22772	Drummond Basin, Qld.	Carboniferous Star of Hope Fm.
p. 79	89	*Lepidodendron sp.*	F 22076	Emerald, Qld.	Late Devonian Telemon Fm.
p. 80	90	*Lepidosigillaria linearis* (Walkom) and *Lepidosigillaria yalwalensis* (Walkom)	AMF 63741	Bunga Head, NSW	Middle Devonian
p. 80	91	*Leptophloeum australe* (McCoy) Walton	AMF 38375	Barraba, NSW	Late Devonian
p. 81	92	*Leptophloeum australe* (McCoy) Walton	F 22072	Emerald, Qld.	Late Devonian, Telemon Fm.
p. 81	93	*Leptophloeum australe* (McCoy) Walton	F 22072	Emerald, Qld.	Late Devonian, Telemon Fm.
p. 81	94	*Leptophloeum australe* (McCoy) Walton	AMF 6502	Goonoo Goonoo, NSW	Late Devonian
p. 83	96	*Adiantites paracasica* Gothan	CPC 4320	Clarke R., Qld.	Early Carboniferous
p. 83	97	*Barinophyton obscurum* (Dun) White	AMF 51179	Genoa R., NSW	Late Devonian
p. 84	98, 99	*Archaeopteris howitti* McCoy	MV p	Iguana Ck., Vic.	Late Devonian
p. 84	100	*Archaeopteris howitti* McCoy	AMF 51185	Genoa R., NSW	Late Devonian
p. 85	101	Unidentified plant	MV p 159791	Howitt's Quarry, Vic.	Late Devonian
p. 89	103	*Rhacopteris ovata* (McCoy) Walkom	AMF 57255	Stroud, NSW	Carboniferous
p. 90	104	*Botrychiopsis plantiana* (Carruthers) Archangelesky & Arrondo	AMF 35633	Patterson, NSW	Carboniferous
p. 91	105	Aphlebium	AMF 58526	Stroud, NSW	Carboniferous
p. 91	106	*Sphenopteridium intermedium* (Feist.) Rigby	AMF 17517	Stroud, NSW	Carboniferous
p. 91	107	*Botrychiopsis plantiana* (Carruthers) Archangelesky & Arrondo	AMF 58524	Raymond Terrace, NSW	Late Carboniferous
p. 92	108	*Sphenopteridium intermedium* (Feist.) Rigby	AMF 57234	Stroud, NSW	Carboniferous
p. 92	109	*Rhacophyllum diversiforme* Eth. fil.	AMF 17521	Stroud, NSW	Carboniferous
p. 92	110	Fruiting bodies of Rhacopterid	AMF 58526	Stroud, NSW	Carboniferous
p. 92	111	*Dactylophyllum digitatum* Eth. fil.	AMF 58526	Stroud, NSW	Carboniferous
p. 93	112	*Subsigillaria sp.*	AMF 57266	Stroud, NSW	Carboniferous
p. 93	113	*Cyclostigma australe* Feist.	AMF 35644	Stroud, NSW	Carboniferous
p. 93	114	Cone	AMF 26929	Clarencetown, NSW	Carboniferous
p. 96	115	Petrified tree trunk	AMF	Werris Ck., NSW	Permian
p. 97	116	*Palaeosmunda williamsii* Gould	AMF 63748	Blackwater, Qld.	Late Permian
p. 98	117	*Glossopteris spp.*	AMF Gallery Display	Newcastle, NSW	Permian
p. 99	118	*Gangamopteris angustifolia* McCoy	MV p 178127	Bacchus Marsh, Vic.	Early Permian
p. 100	119	*Gangamopteris walkomii* Rigby	AMF 43564	Katoomba, NSW	Permian
p. 100	120	*Palaeovittaria sp.*	AMF 25572	Belmont, NSW	Late Permian Insect Beds
p. 100	121	*Gangamopteris cyclopteroides* Feist.	MV p 48499	Bacchus Marsh, Vic.	Early Permian
p. 101	122	*Glossopteris sp.* cuticle	AMF	Richmond Vale, NSW	Permian
p. 101	123	*Glossopteris sp.* and *Noeggerathiopsis hislopi* (Bunbury) Feist.	AMF 66268	Newcastle, NSW	Permian
p. 102	124	*Vertebraria indica* Royle	AMF 63723	Newcastle, NSW	Permian
p. 102	125	*Vertebraria indica* Royle	AMF 11018	Balmain, NSW	Permian
p. 103	126	*Glossopteris sp.*	AMF 25521	Warner's Bay, NSW	Late Permian Insect Beds
p. 103	127	*Glossopteris spp.*	AMF	Cooyal, NSW	Late Permian
p. 104	128	*Glossopteris spp.*	AMF 56472	Dunedoo, NSW	Permian
p. 104	129	*Glossopteris duocaudata* Holmes	AMF 60028	Cooyal, NSW	Permian
p. 104	130	*Glossopteris sp.*	MEW	Dunedoo, NSW	Permian
p. 105	131	*Gangamopteris walkomii* Rigby	AMF 43528	Katoomba, NSW	Permian
p. 105	132	*Glossopteris linearis* (McCoy) White	AMF 48156	Mudgee, NSW	Permian

Page	No.	Name and author of species	Fossil no.	Locality	Age
p. 106	133, 134, 135, 136	*Glossopteris venation*	AMF	Cooyal, NSW	Permian
p. 107	137	*Glossopteris* venation	AMF	Cooyal, NSW	Permian
p. 108	138	*Plumsteadia ampla* (White) Rigby	F22286	Duaringa, Qld.	Permian Upper Bowen CM.
p. 110	139	*Plumsteadia* on *Glossopteris sp.*	AMF 57485	Mudgee, NSW	Late Permian, Illawarra CM
p. 111	140	*Plumsteadia ampla* (White) Rigby	F22297	Duaringa, Qld.	Late Permian
p. 111	141	*Plumsteadia sp.*	AMF 66030	Mudgee, NSW	Permian
p. 111	142	*Plumsteadia ampla* (White) Rigby	F22297	Duaringa, Qld.	Late Permian
p. 112	143	*Dictyopteridium sporiferum* Feist.	AMF 26130	Newcastle, NSW	Permian
p. 112	144	*Dictyopteridium sporiferum* Feist.	AMF 55035	Cooyal, NSW	Permian
p. 113	145	*Scutum sp.* on *Glossopteris sp.*	AMF 66563	La Trobe, Tas.	Permian
p. 114	146, 147	New type of Glossopterid fructification	AMF 19713, 19714	Belmont, NSW	Late Permian Insect Beds
p. 114	148	*Austroglossa walkomii* Holmes	AMF 55027	Cooyal, NSW	Permian
p. 114	149	*Nummulospermum bowense* Walkom	AMF 29758	Newcastle, NSW	Permian
p. 114	150	*Samaropsis sp.* attached to *Glossopteris sp.*	MMF 23231	Newcastle, NSW	Permian
p. 116	151	Fertile Glossopterid scale	MMF 7473	Co. Lincoln, NSW	Permian
p. 116	152	*Partha belmontensis* White	AMF 46525	Belmont, NSW	Late Permian Insect Beds
p. 116	153	Fertile Glossopterid scale-leaf bearing *Indocarpus sp.* seeds	MMF 7472	Co. Lincoln, NSW	Permian
p. 117	154	*Rigbya arberioides* Lacey	AMF 29086	Newcastle, NSW	Permian
p. 117	155	*Rigbya arberioides* Lacey	AMF 57666	Warner's Bay, NSW	Late Permian Insect Beds
p. 117	156	"*Mudgea ranunculoides* Melville"	BMNH	Ulan, NSW	Late Permian
p. 118	157	*Eretmonia natalensis* du Toit	AMF 41553	Belmont, NSW	Late Permian Insect Beds
p. 118	158	*Squamella australis* White	AMF 43706	Belmont, NSW	Late Permian Insect Beds
p. 119	159	*Squamella australis* White	AMF 57626	Tuggerah, NSW	Late Permian
p. 119	160	*Squamella australis* White	AMF 57625, 57626	Tuggerah, NSW	Late Permian
p. 120	161	*Glossopteris linearis* McCoy and *Squamella australis* White	AMF 57342	Newcastle, NSW	Permian
p. 121	162, 163	*Squamella sp.*	MMF 17483	Ulan, NSW	Permian
p. 123	164	*Walkomiella australis* (Feist.) Florin	AMF 35658	Ulan, NSW	Late Permian
p. 123	165, 166	*Walkomiella australis* (Feist.) Florin	AMF 46262	Ulan, NSW	Late Permian
p. 123	167	*Walkomiella australis* (Feist.) Florin	AMF 42263	Ulan, NSW	Late Permian
p. 123	168	*Walkomiella australis* (Feist.) Florin	AMF 46086	Ulan, NSW	Late Permian
p. 126	172	*Blechnoxylon talbragarensis* Eth. fil.	AMF 6300	Talbragar, NSW	Late Permian, Illawarra CM
p. 126	173	*Blechnoxylon talbragarensis* Eth. fil.	AMF 6300	Talbragar, NSW	Late Permian, Illawarra CM
p. 126	174	*Blechnoxylon talbragarensis* Eth. fil.	AMF 6299	Talbragar, NSW	Late Permian, Illawarra CM
p. 127	175	*Blechnoxylon talbragarensis* Eth. fil.	AMF 6299	Talbragar, NSW	Late Permian, Illawarra CM
p. 128	176	*Sphenopteris lobifolia* Morris	AMF 26112	Hamilton, NSW	Permian
p. 128	177	Fertile fern	AMF 66589	Mt. Victoria, NSW	Late Permian
p. 128	178	*Sphenopteris flexuosa* McCoy	AMF 29286	Newcastle, NSW	Permian
p. 128	179	*Alethopteris lindleyana* (Royle) Schimper	AMF 14948	Tryphinia, Qld.	Permian
p. 129	180	*Alethopteris lindleyana* (Royle) Schimper	F22180	Duaringa, Qld.	Late Permian, Upper Bowen CM
p. 130	181	*Phyllotheca australis* Bgt.	AMF 19662	Shepherd's Hill, Newcastle, NSW	Permian
p. 130	182	*Stellotheca robusta* Surange & Prakash	AMF 14948	Tryphinia, Qld.	Permian
p. 131	183	*Umbellaphyllites ivinii* (Walkom) Rigby	AMF 58540	Burning Mt., NSW	Permian
p. 131	184	*Phyllotheca etheridgei* Arber	AMF 50940	Shepherd's Hill, Newcastle, NSW	Permian
p. 132	185	*Noeggerathiopsis hislopi* (Bunbury) Feist.	AMF 41735	Newcastle, NSW	Permian
p. 132	186	Stem with seasonal banding	MMF 3009	Gunnedah, NSW	Permian
p. 132	187	Ginkgophyte leaf	MMF 2996	Cooyal, NSW	Permian
p. 133	188	*Dunedoonia reticulata* Holmes	AMF 56472	Dunedoo, NSW	Permian
p. 133	189	*Samaropsis pincombe* Walkom	AMF 48127	Bar Beach, Newcastle, NSW	Permian

Page	No.	Name and author of species	Fossil no.	Locality	Age
p. 133	190	*Schizoneura gondwanensis* Feist.	AMF 39579	Newcastle, NSW	Permian
p. 137	191	*Cylomeia sp.* rhizophore	AMF 61557	Avalon, NSW	Early Triassic, Narrabeen Gp.
p. 137	192	*Cylomeia sp.*	AMF 68364	Avalon, NSW	Early Triassic, Narrabeen Gp.
p. 138	193	*Cylomeia undulata* (Burges) White	AMF 58791	Bellambi, NSW	Late Permian to Basal Triassic
p. 138	194	*Dicroidium callipteroides* (Carpentier)	AMF 59519	Bulli, NSW	Late Permian to Basal Triassic
p. 138	195	Lycopod	MMF 25687	Bulli, NSW	Late Permian to Basal Triassic
p. 138	196	*Voltziopsis africana* (Seward) Townrow and *Podocarpus sp.*	AMF 22426	Bulli, NSW	Late Permian to Basal Triassic
p. 139	197	*Cladophlebis australis* (Morris) Seward	AMF 60237	Turrimetta Head, NSW	Early Triassic, Narrabeen Gp.
p. 139	198	*Taeniopteris wianamattae* (Feist.)	AMF 57975	Warriewood, NSW	Early Triassic, Narrabeen Gp.
p. 139	199	*Voltziopsis wolganensis* Townrow	AMF 48168	Kanimbla Valley, NSW	Triassic
p. 139	200	*Phyllotheca sp.*	AMF 60818	Avalon, NSW	Early Triassic, Narrabeen Gp.
p. 140	201	*Cylomeia undulata* (Burges) White	AMF 63735	Bulli, NSW	Late Permian to Basal Triassic
p. 140	202	*Cylomeia capillamentum* White	AMF 60890	Narrabeen, NSW	Early Triassic, Narrabeen Gp.
p. 140	203	*Cylostrobus sydneyensis* (Walkom)	AMF 66625	Narrabeen, NSW	Early Triassic, Narrabeen Gp.
p. 140	204	*Cylomeia capillamentum* White Cone	AMF 66625	Narrabeen, NSW	Early Triassic, Narrabeen Gp.
p. 140	205	*Cylomeia sp.*	MMF 25870	Capertee Valley, NSW	Early Triassic
p. 141	206	*Cylomeia longicaulis* leaves and *Cylostrobus sydneyensis* cones	AMF 59975	Long Reef, NSW	Early Triassic, Narrabeen Gp.
p. 141	207	*Cylomeia undulata* (Burges) White	AMF 59987	Narrabeen, NSW	Early Triassic, Narrabeen Gp.
p. 141	208	*Skilliostrobus australis* Ash	AMF 58856	Terrigal, NSW	Triassic
p. 142	209	*Dicroidium zuberi* (Szajnocha) Archangelesky	AMF 44472	Beacon Hill, NSW	Triassic
p. 143	210	*Dicroidium dubium var dubium* (Feist.) Gothan	AMF 61479	Benolong, NSW	Middle Triassic
p. 144	211	*Dicroidium odontopteroides var moltenensis* Retallack	AMF 61131	Benolong, NSW	Middle Triassic
p. 144	212	*Dicroidium sp.*	AMF 45709	Delungra, NSW	Middle Triassic, Gunnee Beds
p. 144	213	*Xylopteris tripinnata* (Jones & de Jersey) Schopf	AMF 61581	Beacon Hill, NSW	Triassic
p. 145	214	*Dicroidium odontopteroides* (Morris) Gothan	F16828	Denmark Hill, Ipswich, Qld.	Triassic
p. 146	215	*Dicroidium odontopteroides*, *Dicroidium elongatum*, and *Linguifolium diemenense* Walkom	AMF 57792	Hobart, Tas.	Triassic
p. 146	216	*Dicroidium odontopteroides* (Morris)	F16828	Denmark Hill, Ipswich, Qld.	Triassic
p. 147	217	*Dicroidium odontopteroides* and *Pterophyllum nathorsti* Seward	MEW	Great Artesian Basin, Qld.	Triassic
p. 147	218	*Dicroidium sp.* cuticle	MEW	South Africa	Triassic
p. 148	219	*Dicroidium* – fertile pinna	AMF 59445	Lorne Basin, NSW	Triassic
p. 148	220	*Pteruchus barrealensis* (Frenguelli) *var feistmanteli* Holmes	AMF 59442	Lorne Basin, NSW	Triassic
p. 148	221	*Umkomasia sp.*	AMF 61487	Benolong, NSW	Middle Triassic
p. 149	222	*Dicroidium* "flowers"	AMF 52170	Clarence Siding, NSW	Middle Triassic
p. 149	223	*Dicroidium* "flowers"	AMF 58795	Mt. Piddington, NSW	Triassic
p. 150	224	Lycopod rhizophore	F23258	Springsure, Qld.	Triassic
p. 150	225	Lycopod stems	CPC 2854	NE Canning Basin, W.A.	Triassic, Erskine Ss.
p. 151	226	*Ginkgoites semirotunda* Holmes	AMF 61512	Benolong, NSW	Middle Triassic
p. 151	227	*Cycadopteris scolopendrina* Ratte	AMF 35645	St Peters, NSW	Triassic, Ashfield Shale
p. 152	228	*Neocalamites hoerensis* (Schimper)	AMF 66329	Beacon Hill, NSW	Triassic

Page	No.	Name and author of species	Fossil no.	Locality	Age
p. 152	229	Equisetalean cone	F23785	Mt Bannerman, Canning Basin, WA	Triassic
p. 152	230	Equisetalean stems	F23767	Canning Basin, WA	Triassic, Erskine Ss.
p. 153	231	*Lepidopteris stormbergensis* (Seward) Townrow	AMF 61701	Benolong, NSW	Middle Triassic
p. 153	232	*Peltaspermum sp.*	AMF 61508	Benolong, NSW	Middle Triassic
p. 154	233	Fertile Fern	AMF 66322	Beacon Hill, NSW	Triassic
p. 154	234	*Rienitsia spatulata* Walkom	AMF 55110	Gosford, NSW	Triassic
p. 155	235	*Todites pattinsoniorum* Holmes	AMF 61119	Benolong, NSW	Middle Triassic
p. 155	236	*Rissikia media* Townrow, and *Phoenicopsis elongatus* (Morris) Seward	AMF 58505	Nymboida, NSW	Triassic
p. 160	238	*Rissikia talbragarensis* White	AMF 59822	Talbragar, NSW	Jurassic
p. 161	239	*Agathis jurassica* White	AMF 59990	Talbragar, NSW	Jurassic
p. 162	240	*Rissikia talbragarensis* White	AMF 38839	Talbragar, NSW	Jurassic
p. 162	241	*Agathis jurassica* cone scale	AMF 61580	Talbragar, NSW	Jurassic
p. 162	242	*Agathis jurassica* – young cone	AMF 59984	Talbragar, NSW	Jurassic
p. 162	243	*Elatocladus planus* (Feist.) Seward	AMF 58289	Talbragar, NSW	Jurassic
p. 163	244	Cone of a Pteridosperm	MMF 8690	Talbragar R., NSW	Jurassic
p. 163	245	*Allocladus cribbi* Townrow	MMF 25832	Talbragar, NSW	Jurassic
p. 163	246	*Agathis jurassica*, *Elatocladus planus*, *Pentoxylon australica* White, *Pachypteris pinnata* (Walkom) Townrow	MMF 28459	Talbragar, NSW	Jurassic
p. 163	247	*Pachypteris crassa* (Halle) Townrow	AMF 26647	Talbragar, NSW	Jurassic
p. 164	248	*Agathis jurassica* White, and *Pentoxylon australica* White	AMF 59990	Talbragar, NSW	Jurassic
p. 165	249	*Pentoxylon australica* White. Fruit.	MMF 3163	Talbragar, NSW	Jurassic
p. 165	250	*Pentoxylon australica* White. Male sporangiophore	AMF 60862	Talbragar, NSW	Jurassic
p. 165	251	*Pentoxylon australica* White. Petrified wood	AMF 63747	Chinchilla, Qld.	Jurassic
p. 166	252	Tree-fern wood	AMF	Tibooburra, NSW	Jurassic
p. 166	253	*Pentoxylon* – petrified wood	AMF 60935	Chinchilla, Qld.	Jurassic
p. 167	254	*Osmundacaulis hoskingii* Gould	AMF	Tibooburra, NSW	Jurassic
p. 167	255	*Pentoxylon australica* White	AMF 63746	Chinchilla, Qld.	Jurassic
p. 168	256	*Aspleniopteris sp.*	CPC 2837	Selwyn, Qld.	Jurassic
p. 169	258	*Phlebopteris alethopteroides* Eth. fil.	AMF 358	Toowoomba, Qld.	Jurassic
p. 169	259	*Isoetes sp.*	MMF 25282	Pilliga, NSW	Jurassic
p. 169	260	*Isoetites elegans* Walkom	AMF 39815	Gingin, WA	Jurassic
p. 170	261	*Pentoxylon australica* White	F23214	Surat Basin	Jurassic, Orallo Fm.
p. 170	262	*Otozamites bechei* Bgt.	AMF 58731	Perth Basin, WA	Jurassic
p. 170	263	*Sagenopteris rhoifolia* Presl.	AMF 35739	Condamine R., Qld.	Jurassic
p. 171	264	*Araucaria feistmanteli* (Halle)	AMF 58731	Yarragadee, WA	Jurassic
p. 171	265	*Bellarinea barkleyi* (McCoy) Florin	AMF 66231	Bex Hill, NSW	Jurassic
p. 171	266	Conifer twig	MEW	Laura Basin, Qld.	Jurassic
p. 177	268	*Rienitsia variabilis* Douglas, and *Rissikia media* (Ten. Woods) Townrow	AMF 57977	Boola Boola, Vic.	Early Cretaceous
p. 178	269	*Sphenopteris sp.*	AMF 68368	Koonwarra, Vic.	Early Cretaceous
p. 179	270	*Pterophyllum fissum* Feist.	F21894	Bauhinia Downs, NT	Early Cretaceous
p. 179	271	Conifer twig	F21894	Bauhinia Downs, NT	Early Cretaceous
p. 179	272	*Microphyllopteris gleichenioides*, Oldham & Morris	F22086	Bauhinia Downs, NT	Early Cretaceous
p. 179	273	*Otozamites bechei* Bgt.	F21891	Bauhinia Downs, NT	Early Cretaceous
p. 180	274	*Araucarites sp.*	AMF 68370	Koonwarra, Vic.	Early Cretaceous
p. 180	275	*Amanda floribunda* Douglas, and *Phyllopteroides dentata* Medwell	AMF 68372	Devil's Kitchen, Vic.	Early Cretaceous
p. 180	276	*Taeniopteris daintreei* McCoy	AMF 68369	Koonwarra, Vic.	Early Cretaceous
p. 180	277	Petal-like organ	AMF 68365	Koonwarra, Vic.	Early Cretaceous
p. 180	278	*Phyllopteroides lanceolata* (Walkom) Medwell	F22959	Quilpie, Qld.	Early Cretaceous
p. 181	279	*Ginkgoites australis* (McCoy) Florin	AMF 68366	Koonwarra, Vic.	Early Cretaceous
p. 181	280	*Phyllopteroides dentata* Medwell	AMF 68371	Devil's Kitchen, Vic.	Early Cretaceous

p. 182	281	*Williamsonia sp.*	F21889	Bauhinia Downs, NT	Early Cretaceous
p. 182	282	*Williamsonia sp.*	F21898	Bauhinia Downs, NT	Early Cretaceous
p. 184	284	*Rissikia media* Townrow	AMF 57977	Boola Boola, Vic.	Early Cretaceous
p. 184	285	*Sphenopteris warragulensis* McCoy	AMF 68367	Koonwarra, Vic.	Early Cretaceous
p. 185	286	*Otozamites bengalensis* (Morris)	F21894	Bauhinia Downs, NT	Early Cretaceous
p. 220	343	*Brachychiton sp.*	AMF 66350	Vegetable Ck., NSW	Tertiary
p. 220	.344	*Bombax sturtii* Ett., and *Magnolia brownii* Ett.	AMF 51329	Dalton, NSW	Tertiary
p. 221	346	*Lomatia brownii* Ett.	AMF 51154	Vegetable Ck. NSW	Tertiary
p. 222	347	*Lomatia* sp.	F13008	Dalton, NSW	Tertiary
	349	*Palaeosmunda sp.*			
	350	*Palaeosmunda sp.*	RS	Lune River, Tas.	Tertiary
	351	Rachis			
	352	*Angiopteris*-like Tree-fern	RS	Lune River, Tas.	Tertiary
p. 223	353 354	*Angiopteris*-like Tree-fern	RS	Lune River, Tas.	Tertiary
p. 224	356	*Hakea duttoni* Ett.	AMF 51159	Vegetable Ck., NSW	Tertiary
p. 225	357	*Fagus hookeri* Ett.	AMF 51156	Vegetable Ck., NSW	Tertiary
p. 225	358	*Roupala sapindifolia* Ett.	AMF 66340	Vegetable Ck., NSW	Tertiary
p. 225	359	*Piper feistmanteli* Ett.	AMF 51125	Penrose, NSW	Tertiary
p. 225	360	*Quercus wilkinsoni* Ett.	AMF 51221	Vegetable Ck., NSW	Tertiary
p. 225	361	*Quercus drymejioides* Ett.	AMF 51128	Dalton, NSW	Tertiary
p. 226	362	*Araucaria crassa* (Ten. Woods) Townrow	AMF	Ipswich, Qld.	Tertiary
p. 227	363	*Phyllocladus asplenioides* Ett.	AMF 51133	Vegetable Ck., NSW	Tertiary
p. 227	364	Angiosperm leaf – Dicot.	AMF 44114	Penrose, NSW	Tertiary
p. 226	365	*Podocarpus cupressinoides* (Ett.) Selling	AMF 44096	Penrose, NSW	Tertiary
p. 226	366	*Cinnamomum sp.*	AMF 44155	Penrose, NSW	Tertiary
p. 226	367	*Cinnamomum leichardti* Ett.	AMF 51122	Elsmore, NSW	Tertiary
p. 227	368	*Elaeocarpus muelleri* Ett.	AMF 31156	Murrurundi, NSW	Tertiary
p. 227	369	Angiosperm leaf. Dicot.	F23615	Marawaka, PNG	Tertiary
p. 232	370	*Angiopteris*-like Tree-fern	RS	Lune River, Tas.	Tertiary

INDEX